AF686224

Table of Contents

The Forgotten Innovators

SANDEEP CHAVAN

Published by SANDEEP CHAVAN, 2025.

While every precaution has been taken in the preparation of this book, the publisher assumes no responsibility for errors or omissions, or for damages resulting from the use of the information contained herein.

THE FORGOTTEN INNOVATORS

First edition. February 6, 2025.

Copyright © 2025 SANDEEP CHAVAN.

ISBN: 979-8230836797

Written by SANDEEP CHAVAN.

To the Independent Researchers of India—

The thinkers, innovators, and visionaries who **pursue knowledge against all odds**,Who challenge the status quo **without institutional backing**,Who refuse to let **bureaucracy, funding constraints, or lack of recognition**deter them from exploring the unknown.

This book is for **every mind that dares to ask questions,every scholar who works in isolation yet dreams of discovery,and every innovator who believes that knowledge should be free and accessible to all.**

May your perseverance shape the future of science,and may your contributions never be forgotten.

"All great truths begin as blasphemies."— George Bernard Shaw

"New ideas pass through three phases: first, they are ridiculed; second, they are violently opposed; third, they are accepted as self-evident."— Arthur Schopenhauer

"The important thing is not to stop questioning. Curiosity has its own reason for existing."— Albert Einstein

These words echo the struggles and triumphs of **independent researchers**—those who **challenge conventions, persist despite rejection, and push the boundaries of knowledge without institutional privilege.**

Preface

Scientific inquiry is the foundation of human progress. Throughout history, curiosity, independent thought, and relentless questioning have driven some of the greatest discoveries—from the laws of motion to quantum mechanics, from medical breakthroughs to artificial intelligence. Yet, in the modern era, where institutional frameworks dictate the direction of research, independent researchers are often overlooked, underfunded, and marginalized.

This book is a critique of the systemic barriers that hinder non-affiliated scientists and thinkers from contributing to innovation and discovery. It highlights the monopolization of research funding by elite institutions, the bias of academic publishing, the bureaucracy that stifles scientific freedom, and the neglect of industry toward fundamental research. As someone who has personally faced the challenges of being an independent researcher, I felt a deep need to document these issues, explore alternative solutions, and advocate for change.

The objective of this book is twofold:

- To expose the challenges that independent researchers face—from the difficulty of securing funding to the lack of lab access and the struggles of publishing groundbreaking work.
- To propose actionable solutions—by exploring global best practices, alternative funding mechanisms, and policy reforms that can make research more inclusive, diverse, and innovation-driven.

India, despite its rich scientific heritage and intellectual capital, remains far behind in supporting independent researchers. Many brilliant minds either migrate abroad in search of better opportunities or abandon their research ambitions due to systemic roadblocks. If this continues, India will remain a consumer of global innovations rather than a leader in scientific breakthroughs.

Who This Book Is For

This book is for:

- Independent researchers struggling to navigate the funding maze, infrastructure constraints, and institutional gatekeeping.
- Policymakers who can influence change by introducing direct research grants, opening public labs to independent scholars, and promoting open-access publishing.
- Industry leaders who have the power to fund high-risk, high-reward fundamental research rather than merely importing technology.
- Academics and students who believe in scientific inquiry beyond the walls of universities and corporate R&D labs.

A Call for Change

The next decade must see a transformation in research policies—not just in India, but globally. The future of knowledge creation should not be confined to institutions alone; it should be accessible to all capable minds willing to contribute to science, technology, and human advancement.

This book is not just a critique but a call for reform. It is an attempt to ignite discussions, push for policy shifts, and inspire a movement that recognizes the importance of independent scientific thought.

To every independent thinker, researcher, and innovator reading this—your work matters. Your ideas deserve recognition. And together, we can work toward a world where scientific progress is driven by curiosity and merit, not bureaucracy and institutional privilege.

The time for change is now.

1. Introduction: The Neglect of Independent Research in India

What is Independent Research, and Why is it Important?

Independent research refers to scholarly or scientific work conducted without direct affiliation with government institutions, universities, or corporate R&D centers. It is driven by individual curiosity, intellectual rigor, and a commitment to knowledge advancement. Independent researchers challenge mainstream perspectives, explore unconventional ideas, and contribute to scientific progress without being confined by institutional limitations.

Globally, independent research has led to groundbreaking discoveries. Albert Einstein formulated the theory of relativity while working as a patent clerk, and Gregor Mendel conducted genetic research in a monastery. Many revolutionary ideas emerged outside formal institutions, proving that research does not always require an institutional framework—what it needs is a conducive ecosystem.

However, in India, independent research is neither encouraged nor supported. The system operates under the assumption that innovation is exclusive to government-backed institutions or corporate-funded R&D. This restrictive approach has crippled individual innovation, forcing researchers either to abandon their pursuits or seek opportunities abroad.

India's Rich Historical Contributions to Science vs. Its Current Failure

Historically, India has been a global center for scientific and philosophical thought. Ancient Indian scholars made significant contributions to mathematics, astronomy, medicine, and metallurgy. Aryabhata's work on astronomy, Sushruta's advancements in surgery, and Bhaskaracharya's theories in calculus predated similar discoveries in the West. Indian innovation thrived independently, without rigid institutional structures.

Post-independence, however, India's scientific priorities shifted towards centralized research organizations, such as CSIR, ISRO, DRDO, and IITs. While these institutions have played a crucial role in advancing applied sciences, they have monopolized the research ecosystem, sidelining independent thinkers. Unlike the past, where research flourished in diverse intellectual settings, today's India restricts innovation to a bureaucratic structure controlled by institutional hierarchies.

The irony is that the government celebrates ancient Indian achievements but fails to create a modern research framework that empowers individual innovators. This contradiction between glorifying past discoveries and neglecting present-day independent researchers reflects a deep flaw in India's scientific policies.

The Invisible Struggle of Self-Funded Researchers

For independent researchers in India, the challenges are immense and often insurmountable. The system provides no formal grants, no access to research labs, no recognition, and no institutional backing. Unlike in the West, where platforms like National Science Foundation (NSF) in the US or Max Planck Society in Germany provide funding for independent researchers, India offers no equivalent mechanism.

Some key struggles include:

- **Financial Constraints:** Independent researchers must self-fund their work, often at personal risk. There are no dedicated grants for individuals not affiliated with institutions.
- **Lack of Research Infrastructure:** Access to high-end labs, advanced computing resources, or even basic research tools is nearly impossible without institutional affiliation.
- **Publishing Barriers:** Prestigious journals prefer institutional affiliations, making it difficult for independent researchers to publish credible work.
- **Government Apathy:** Indian policymakers do not recognize independent research as a legitimate contributor to national scientific progress. There are no clear policies to support or integrate independent researchers into the mainstream.

The neglect of independent research is not just a policy failure; it is a national loss. India's greatest untapped intellectual resource remains ignored, leading to brain drain and intellectual stagnation.

Conclusion

If India aspires to be a global knowledge hub, it must recognize and support independent researchers. The current system, which favors institutional hierarchy over individual brilliance, must be reformed. Without a fundamental shift in policies, India will continue to lose its brightest minds to more supportive research ecosystems abroad.

2: The Government's Role – Apathy, Bureaucracy, and Red Tape

Independent research in India does not suffer from a lack of talent, ideas, or dedication—it suffers from systemic neglect. The government, which should ideally nurture innovation across all spectrums, has instead created a bureaucratic fortress where research funding, resources, and recognition are reserved exclusively for institutional affiliations. Independent researchers are left invisible to policymakers, trapped in a system where only universities, IITs, IIMs, and government-funded laboratories have access to grants and infrastructure.

Even when funding is theoretically available, layers of red tape, opaque application processes, and favoritism make it nearly impossible for independent researchers to secure financial support. Unlike global models that encourage both institutional and independent research, India's framework remains rigid, discouraging self-driven innovators from contributing to the nation's knowledge economy.

This chapter critically examines how government policies actively exclude independent researchers, how the bureaucratic funding system obstructs innovation, and why India lags behind other countries in providing structured support for non-institutional researchers.

2.1 Institutional Monopoly: How Government Funding Policies Favor Institutions Over Individuals

The Indian research ecosystem operates under a deep-rooted institutional monopoly, where government funding is reserved primarily for universities, IITs, NITs, and government-backed research labs. While these institutions undoubtedly contribute to scientific progress, the system completely disregards independent researchers, who remain excluded from formal funding channels. This institutional bias creates a rigid hierarchy where innovation is expected to emerge only within predefined structures, leaving independent thinkers without the resources necessary to pursue their work.

The Bias Toward Universities, IITs, and Government-Backed Labs in Research Grants

Most government-funded research grants in India are controlled by bodies such as:

- Science and Engineering Research Board (SERB)
- Department of Science and Technology (DST)
- Council of Scientific and Industrial Research (CSIR)
- Indian Council of Medical Research (ICMR)
- Defence Research and Development Organisation (DRDO)

Each of these organizations has funding schemes designed exclusively for researchers affiliated with recognized institutions. Independent researchers, no matter how qualified or innovative, do not qualify for these grants unless they are attached to a university or government-funded lab.

This preference for institutional affiliation over individual merit leads to several fundamental problems:

- **Discouragement of Non-Traditional Innovators:** Many breakthroughs have historically come from individuals outside formal institutions (e.g., Einstein, Tesla, and Ramanujan). By restricting funding to institutional researchers, India prevents self-driven innovators from contributing to national progress.
- **Bureaucratic Restrictions on Idea-Driven Research:** Most university-based research projects are rigidly structured around faculty interests and institutional priorities. Independent researchers, who often explore unconventional ideas, do not fit into this model and are consequently excluded.
- **Prioritization of Prestige Over Potential:** Many funding agencies prefer high-ranking institutions (IITs, IISc, IISERs) over smaller colleges or independent scholars, assuming that quality research only emerges from reputed organizations. This elitist funding approach discourages grassroots innovation.

The Eligibility Barriers That Exclude Independent Researchers From National Funding Schemes

Independent researchers in India face multiple layers of exclusion when applying for grants:

- **Institutional Affiliation Requirement:** Most research grants require applicants to be permanent faculty members or PhD students at a recognized institution. This immediately eliminates independent scholars who work outside of formal academic settings.
- **Lack of Infrastructure Support:** Even if an independent researcher has a groundbreaking idea, the absence of access to laboratories, equipment, or research databases prevents them

from conducting experiments or publishing their findings.

- **Complicated Administrative Procedures:** Applying for government grants requires navigating complex bureaucratic procedures, including approvals from research committees, ethical review boards, and institutional heads—all of which are inaccessible to independent researchers.
- **Preference for Established Research Agendas:** Funding bodies often prioritize mainstream research areas (aligned with government or corporate interests), making it difficult for independent researchers exploring emerging fields to secure funding.

These barriers ensure that only institutional researchers can access financial support, effectively blocking independent scholars from contributing to national innovation.

Case Studies of Independent Researchers Struggling to Secure Funding in India

The real-world struggles of independent researchers highlight the flaws in India's research funding model. Here are some cases that exemplify the systemic exclusion faced by non-affiliated researchers:

Case 1: The Inventor Who Couldn't Patent His Work

Dr. Raghav, an independent mechanical engineer, developed an innovative water purification system using nanotechnology. Despite multiple attempts to secure government research grants, he was repeatedly denied funding due to a lack of institutional affiliation. He eventually tried to file a patent independently, but due to financial constraints and the lack of support, his research remained unpublished. Meanwhile, similar research in IITs received substantial government funding and corporate sponsorships.

Case 2: The Self-Taught Biologist Who Had No Lab Access

Anita, a self-taught researcher in microbiology, discovered a potential organic pesticide alternative that could significantly reduce chemical usage in agriculture. However, since she was not associated with any university or research lab, she was denied access to advanced testing facilities. Without an institution to support her, she had to rely on personal savings and informal lab arrangements, significantly delaying her research.

Case 3: The AI Researcher Ignored by National Funding Bodies

Vikram, an independent AI and machine learning researcher, developed an algorithm that could enhance cybersecurity in financial transactions. He applied for government funding under AI research programs, but his application was rejected because he did not belong to a recognized university or an IIT research lab. Later, he was approached by a foreign tech company that offered him full funding and a research position, forcing him to take his work abroad.

These cases represent just a fraction of the talent India loses every year due to an exclusionary funding system. While the government frequently laments the brain drain, it continues to ignore the talent within the country by denying resources to independent researchers.

Conclusion

The institutional monopoly over research funding in India is a critical roadblock to independent innovation. By tying funding exclusively to universities and government labs, the system prevents individual researchers from making meaningful contributions. The eligibility barriers, bureaucratic red tape, and bias toward prestigious institutions ensure that only a select few receive funding, while independent researchers are left to struggle without support.

If India aspires to be a global leader in innovation, it must revise its research funding policies to support independent researchers. Otherwise, the nation will continue losing its brightest minds to foreign institutions that offer them the recognition and resources they deserve.

2.2 The Grant Application Maze: Complexity, Delays, and Bureaucratic Hurdles

While securing research funding is a challenge worldwide, in India, it is a bureaucratic nightmare—especially for independent researchers. The grant application process is designed exclusively for institutional researchers, making it inaccessible to independent scholars. Even those who meet the eligibility criteria face multiple layers of approvals, unnecessary administrative delays, and favoritism, ensuring that only a select few receive funding. This deeply flawed system discourages genuine research efforts, delays innovation, and forces many researchers to either self-fund or abandon their projects entirely.

How Grant Applications Are Designed for Institutional Researchers, Making Them Inaccessible for Independent Scholars

Government research grants in India follow a rigid structure, assuming that all researchers belong to universities or government-funded labs. This creates an immediate exclusion for independent researchers who do not fit into these predefined categories. Some key structural flaws include:

1. Institutional Sponsorship Requirement

- Most major funding agencies—including DST, SERB, CSIR, and ICMR—require applicants to be affiliated with recognized research institutions.
- Independent researchers are expected to secure a host institution before applying, which is nearly impossible unless they have personal connections in academia.
- Many universities refuse to sponsor external researchers because it does not align with their internal funding priorities.

2. Strict Eligibility Criteria

- Grants often specify that applicants must be full-time faculty members, doctoral students, or post-doctoral fellows, eliminating the possibility of independent scholars applying.
- Even well-qualified researchers with published work and patents struggle to obtain funding because they do not have a formal academic title.

3. Lack of Alternative Funding Streams for Individuals

- Unlike countries such as Germany and the United States, where independent researchers can apply for government fellowships or private research grants, India offers no structured funding options for individuals.
- The absence of independent research fellowships forces self-funded researchers to rely on personal savings or foreign grants.

The Multi-Layered Approval System, Which Delays Funding and Discourages Applicants

Even for those who qualify, the application and approval process is so slow and inefficient that it often takes years to receive funding—by which time research priorities may have changed, making the funding irrelevant. Some of the biggest obstacles include:

1. Multiple Bureaucratic Approvals

- Grant proposals must pass through several levels of review, including institutional research committees, government panels, and ministry approvals.
- Each stage involves extensive documentation, verification, and compliance checks, creating unnecessary delays.
- Many applications get stuck in bureaucratic loops, with

officials asking for irrelevant modifications just to process the paperwork.

2. Excessive Paperwork and Compliance Burdens

- Indian funding agencies require dozens of supporting documents, from detailed project proposals to institutional endorsements and financial guarantees.
- Even minor clerical errors can result in months-long rejections, forcing researchers to restart the process from scratch.
- In contrast, global research grants (such as those in the EU and US) often rely on simple online submissions with fast-tracked approvals for high-potential projects.

3. Delayed Disbursement of Funds

- Even after grants are approved, funds are released in multiple installments with each phase requiring additional approvals.
- Researchers often experience long gaps between approvals and fund disbursement, making it difficult to sustain research continuity.
- Unlike in Western nations, where funds are granted upfront, Indian funding agencies micromanage expenditures, slowing down progress.

The Role of Political and Institutional Favoritism in Determining Grant Approvals

The final and perhaps most damaging aspect of the research funding system in India is favoritism and nepotism. Instead of being awarded based on merit and research potential, grants are often allocated based on personal connections, institutional prestige, or political influence.

1. Preference for Established Institutions Over Merit

- Research grants are disproportionately awarded to IITs, IISc, and select central universities, leaving lesser-known institutions and independent researchers with almost no access to funding.
- Even within these institutions, grants are frequently given to faculty members with strong political or administrative connections, sidelining more deserving candidates.

2. Political Influence in Research Funding

- Many government grants are tied to political agendas, with funding agencies prioritizing government-driven research initiatives over truly independent studies.
- Researchers working on controversial or unconventional topics—such as climate change policies, AI ethics, or political science—often find it impossible to secure funding if their work does not align with government narratives.

3. Opaque Selection Processes

- Grant selection committees often operate without transparency, making it difficult to understand why certain proposals are approved while others are rejected.
- There is no formal grievance redressal mechanism for applicants who believe they were unfairly denied funding.

Conclusion

India's grant application process is deeply flawed, discouraging genuine research efforts and forcing many researchers out of the system. Instead of supporting groundbreaking independent research, the government funding structure is designed to serve institutions and bureaucratic interests.

Unless the system is reformed to be more inclusive, transparent, and efficient, India will continue to lose its brightest minds to international institutions that offer them the support and recognition they deserve.

2.3 The Missing Incentives: Why Independent Research is Not Encouraged

While institutional researchers in India receive structured funding, fellowships, and infrastructure support, independent researchers face a total absence of incentives. The lack of financial aid, mentorship, and tax benefits makes it nearly impossible for self-driven researchers to sustain their work. This systematic neglect has created a hostile environment where independent research is not just difficult but actively discouraged.

Lack of Financial Aid, Fellowships, or Scholarships for Independent Researchers

Globally, independent researchers often have access to grants, fellowships, and research scholarships provided by governments, private institutions, or philanthropic organizations. In India, however, the funding ecosystem is rigidly confined to institutional affiliations, leaving independent researchers with no structured financial aid.

1. No Dedicated Grants for Independent Researchers

- Government research grants are tied to universities, IITs, and research institutes, effectively blocking independent scholars from applying.
- Even private research grants in India are usually routed through academic institutions, making it impossible for independent researchers to access them.
- International funding bodies often require government

approvals for research grants, which independent researchers struggle to obtain due to bureaucratic hurdles.

2. The Fellowship Barrier

- Indian research fellowships, such as UGC NET-JRF, CSIR Fellowship, and ICMR Fellowships, are exclusively for institutional scholars.
- Many international fellowships (e.g., Rhodes, Fulbright, Humboldt Fellowships) require applicants to have academic endorsements from recognized institutions, again excluding independent researchers.
- Without financial assistance, independent researchers must self-fund their studies, a model that is neither sustainable nor productive.

3. Private Sector Apathy Towards Individual Researchers

- In countries like the US and Germany, corporations and private donors support independent researchers through innovation grants and startup funding.
- Indian private sector giants, however, invest minimally in research and development and rarely support independent research unless there is a direct commercial benefit.
- Without government incentives or policy frameworks encouraging private funding for independent research, individual researchers are left with no financial backing.

No Structured Mentorship Programs or Collaboration Platforms for Non-Affiliated Researchers

In academia, mentorship plays a crucial role in guiding researchers, refining ideas, and opening collaboration opportunities. However, for independent researchers in India, there is no structured mentorship framework to support their intellectual growth or professional networking.

1. Absence of Government-Backed Research Networks

- Countries like Germany (Max Planck Society) and the US (National Science Foundation) provide independent researchers access to mentorship programs, research networks, and collaboration hubs.
- India lacks such platforms, making independent researchers feel isolated and disconnected from the broader academic community.

2. No Collaboration Opportunities with Institutional Researchers

- Institutional researchers (e.g., faculty in IITs, IISc) are rarely encouraged to collaborate with independent scholars, as funding structures and institutional policies do not recognize non-affiliated contributors.
- In contrast, countries like the US, UK, and Canada promote open research collaborations, where independent researchers can work alongside university scholars without formal institutional ties.

3. The Lack of Open Research Forums

- There are few or no open-access research incubators in India where independent scholars can exchange ideas, collaborate, and access research materials.
- The absence of government-supported research discussion platforms further isolates independent researchers, reducing their chances of gaining recognition or funding.

The Absence of Tax Benefits or Subsidies for Self-Funded Research Initiatives

One of the most discouraging aspects of independent research in India is the lack of financial incentives in the form of tax breaks, subsidies, or research allowances. In many developed countries, independent researchers receive tax benefits and financial support, but India offers no such relief.

1. No Tax Deductions on Research Expenses

- Countries like the US and UK allow independent researchers to claim tax deductions on research-related expenses such as books, lab equipment, and travel.
- India, however, offers no tax relief, meaning independent researchers must bear 100% of their research costs with no financial reprieve.

2. No Government Support for Self-Funded Research

- In China, researchers conducting self-funded studies can apply for government reimbursements if their work contributes to national development.
- In contrast, India has no reimbursement model, forcing self-funded researchers to either abandon their projects or seek foreign funding.

3. Research Patents and Startups Receive No Incentives

- In countries like Germany and Israel, independent researchers who develop patents, innovations, or research startups receive tax incentives and startup grants.
- In India, the patent process is expensive and complicated, and self-funded researchers receive no financial support to commercialize their work.

Conclusion

The Indian research system is structurally hostile to independent scholars, providing no financial aid, no mentorship, and no tax relief. Without institutional affiliation, independent researchers are left in a vacuum, forced to self-finance their work or give up entirely.

If India is serious about fostering innovation, it must reform its policies to include:

- Dedicated research fellowships for independent researchers.
- Government-backed mentorship and collaboration networks.
- Tax benefits and financial subsidies for self-funded research.

Until such reforms are introduced, independent research in India will remain an uphill battle, forcing many of the country's brightest minds to either struggle in isolation or take their talent elsewhere.

2.4 Global Perspective: How Other Countries Support Independent Researchers

While India systematically neglects independent researchers, other countries have established funding mechanisms, research networks, and institutional support to encourage individual innovation. Nations like the United States, Germany, and China recognize that scientific progress is not confined to institutions and have developed models that allow both institutional and non-institutional researchers to contribute meaningfully to national and global advancements.

By examining these successful models, India can understand the gaps in its research ecosystem and implement policy changes to foster independent research.

United States: National Science Foundation (NSF) and Private Grants for Individual Researchers

The United States is one of the best countries for independent researchers, offering multiple funding sources, tax incentives, and private grants.

1. National Science Foundation (NSF) Grants for Independent Researchers

- The NSF provides direct research grants to independent scholars, startups, and even non-profit research institutions.
- Many NSF grants do not require institutional affiliation, allowing individual researchers to apply based on merit and project feasibility.
- Unlike in India, where applications are trapped in bureaucratic delays, NSF follows a transparent and time-bound approval process.

2. Private Sector and Philanthropic Support

- Large tech companies (such as Google, Microsoft, and IBM) offer innovation grants and research fellowships to independent scholars.
- Organizations like the Howard Hughes Medical Institute (HHMI) and the Gates Foundation fund individual researchers, especially in healthcare and AI.
- Crowdfunding platforms such as Experiment.com enable independent researchers to raise funds from the public, something that is virtually non-existent in India.

3. Government Support for Patents and Commercialization

- The US government offers tax breaks and incentives to independent researchers who develop patentable innovations or tech startups.
- Research-backed startups receive funding through programs

such as the Small Business Innovation Research (SBIR) program, which supports independent innovators without institutional backing.

The US model proves that governments do not need to control all research funding—they can instead empower individual researchers by providing direct grants, private funding opportunities, and commercialization support.

Germany: Max Planck Society's Support for Non-Institutional Scientists

Germany has one of the most well-structured independent research ecosystems, primarily supported by the Max Planck Society and the Deutsche Forschungsgemeinschaft (DFG).

1. The Max Planck Society's Independent Research Model

- Unlike India's institutionally locked research system, the Max Planck Society allows researchers to work independently without needing a university affiliation.
- Independent researchers can access world-class labs, research facilities, and funding if they demonstrate potential for high-impact research.
- The Max Planck Institutes focus on basic research, ensuring that independent scholars can pursue fundamental scientific inquiries without commercial pressure.

2. The Deutsche Forschungsgemeinschaft (DFG) Grants

- The DFG funds independent researchers through fellowships and project-based grants, promoting autonomous research culture.
- Unlike Indian grants, which require institutional approval, DFG grants are reviewed purely on scientific merit.

3. Public and Industry Collaboration in Research

- Germany has well-integrated industry-academia collaboration, but independent researchers are also part of this network.
- Companies such as Siemens, Bosch, and Volkswagen fund independent researchers to work on specialized problems.

The German model shows that a country does not need to rely solely on institutional research—it can foster innovation by investing in independent scholars and providing them direct access to resources.

China: Government-Backed Innovation Funds Open to Both Institutional and Independent Researchers

China has rapidly expanded its research ecosystem, actively funding both institutional and independent researchers. The government treats research as a strategic priority, making significant investments in open research programs.

1. Direct Government Funding for Independent Research

- The National Natural Science Foundation of China (NSFC) provides grants not only to university researchers but also to independent scientists and small research teams.
- Funding is not restricted to large institutions, allowing individual researchers to contribute to key scientific advancements.

2. State-Supported Innovation and Startup Grants

- Independent researchers in China can apply for state-backed innovation funds, particularly in emerging fields such as AI, quantum computing, and biotechnology.
- Unlike in India, where researchers must secure corporate backing, China provides direct funding for independent

innovation through programs like the Torch Program and High-Tech Research & Development Plan (863 Program).

3. Infrastructure Access for Independent Researchers

- Independent researchers in China can access national research facilities, AI labs, and biotechnology research centers at subsidized costs or for free.
- The Chinese government actively encourages non-affiliated scientists to contribute to national research goals, rather than blocking them from the system.

China's model shows that a strong government-backed system can support independent researchers without being restrictive or exclusionary.

Lessons India Can Learn from Global Best Practices in Research Funding

India's research policies need urgent reforms to prevent further brain drain and intellectual stagnation. Based on global best practices, the following changes are necessary:

1. Create Direct Research Grants for Independent Scholars

- India must introduce research fellowships and direct grants for independent researchers, similar to NSF in the US or DFG in Germany.
- Government bodies should allow non-affiliated researchers to apply for research funding without institutional sponsorship.

2. Establish Open Research Infrastructure

- Independent researchers should be given access to government labs, scientific databases, and high-performance computing resources.

- A national Research Infrastructure Sharing Policy should be developed to enable collaborative research between institutions and independent scientists.

3. Encourage Private and Crowdsourced Funding for Research

- India needs to create private-sector research funding models similar to the US and Germany, where corporations invest in independent researchers.
- Platforms like Experiment.com should be developed for public-funded research initiatives.

4. Provide Tax Benefits and Incentives for Independent Researchers

- Independent researchers should be eligible for tax deductions on research expenses such as lab equipment, journal subscriptions, and data collection.
- Startups emerging from independent research should receive government-backed funding and commercialization support.

5. Reduce Bureaucracy and Improve Transparency in Research Funding

- Research grants should have a clear, transparent selection process, preventing favoritism and institutional bias.
- The government should establish an independent research funding body, separate from institutional control, to ensure fair grant distribution.

Conclusion

India's failure to support independent researchers is not due to a lack of resources but due to a lack of political will and policy reform.

Countries like the US, Germany, and China have successfully integrated independent research into their national innovation strategies, allowing non-institutional scholars to thrive.

If India wants to compete globally in innovation and scientific leadership, it must adopt a more inclusive approach by:

- Providing direct funding for independent researchers
- Offering access to research infrastructure
- Creating private-sector and government-backed funding opportunities
- Removing bureaucratic barriers that discourage independent innovation

Until such reforms are implemented, India will continue to lose its brightest minds to countries that recognize and nurture independent research.

2.5 The Consequences of Neglect: How Bureaucracy is Killing Innovation

The systemic neglect of independent researchers in India has far-reaching consequences. It has not only forced many of the country's brightest minds to leave for foreign institutions but has also stifled innovation, discouraged intellectual risk-taking, and weakened India's global standing in research and development. Instead of being an engine of discovery, India's bureaucratic and restrictive research ecosystem acts as a barrier to innovation, preventing potentially transformative ideas from materializing.

2.5.1 How Talented Indian Researchers Migrate Abroad for Better Funding Opportunities

The lack of support for independent researchers in India has led to a mass exodus of scientific talent, commonly referred to as the brain drain. The rigid, institutionally biased funding system has made it nearly impossible for brilliant researchers with unconventional ideas to secure funding, publish, or gain access to world-class facilities. As a result, many scientists and researchers leave India in search of better opportunities in the US, Europe, and China, where governments and private organizations actively encourage independent research and innovation.

1. The Numbers Behind India's Brain Drain

- According to UNESCO data, more than 300,000 Indian students and researchers leave the country every year for higher education and research.
- A study by the National Science Foundation (NSF) in the US found that over 75% of Indian PhD holders in science and engineering prefer to stay abroad due to better research funding and career opportunities.
- Over the past decade, countries like the US, Canada, and Germany have increased research fellowships for Indian scientists, attracting top talent.

2. The Attraction of Foreign Research Ecosystems

Unlike India, where research grants are tied to institutional affiliations, many foreign countries offer:

- Open research fellowships for independent scientists.
- Private and government-backed innovation grants for early-stage research.
- Fast-tracked funding approvals without bureaucratic delays.
- Better salaries and research incentives for individuals pursuing non-traditional projects.

3. Case Studies: Indian Researchers Who Thrived Abroad

- Venkatraman Ramakrishnan, a Nobel Prize-winning

structural biologist, had to conduct his research in the US and UK because India lacked the necessary support for independent structural biology research.

- Abhijit Banerjee, Nobel laureate in economics, spent most of his research career abroad due to limited research funding and institutional constraints in India.
- Many top AI, biotech, and space technology researchers of Indian origin are now working in NASA, MIT, Harvard, and Stanford, instead of contributing to Indian R&D.

The consequence is clear: India is losing its best researchers to countries that provide them the freedom and funding they need to innovate.

2.5.2 The Stifling Effect of Excessive Red Tape on Breakthrough Ideas

The Indian research ecosystem is plagued by excessive bureaucracy, where securing funding often becomes a battle against administrative inefficiencies rather than a merit-based process. While research should thrive in an environment of freedom and open inquiry, Indian researchers must instead navigate layers of approvals, compliance requirements, and institutional roadblocks before they can even begin their work.

1. The Endless Grant Approval Process

Unlike in the US and Germany, where researchers can apply for individual grants with minimal paperwork, Indian researchers must submit applications that require:

- Institutional endorsements.
- Multiple government approvals.
- Clearance from bureaucratic committees that lack technical expertise.
- Ethical reviews that often delay projects by months or even

years.

Many researchers abandon their projects because the time taken to secure funding is longer than the duration of the research itself.

2. Rigid Research Priorities That Discourage Unconventional Ideas

- India's research funding is often dictated by government priorities, leaving little room for exploratory or high-risk innovation.
- Unlike in China and the US, where independent researchers can pursue breakthrough projects in AI, quantum computing, or renewable energy, Indian research funds are largely directed towards applied sciences with immediate commercial benefits.

Many of the most innovative scientific breakthroughs in history, such as quantum mechanics, the theory of relativity, or the structure of DNA, were the result of unconventional, curiosity-driven research—an approach that India's bureaucratic research system actively discourages.

3. How Bureaucratic Delays Killed Promising Indian Research Projects

- Dr. Ramesh, a biophysicist, spent three years waiting for research funding approvals to test a new diagnostic tool for cancer detection. By the time he received partial funding, a US-based team had already patented a similar discovery.
- A team of AI researchers from Bengaluru applied for funding to develop an Indian-language voice recognition model, only to be stuck in approvals for two years. Meanwhile, Google and Microsoft launched their own Indian-language AI models.
- Space researchers working on satellite propulsion technology faced such delays in funding and resource allocation that they

eventually moved to NASA and ESA, where their work was fast-tracked.

These examples highlight how excessive bureaucracy actively obstructs Indian innovation. If India does not reform its research funding system, it will continue to fall behind in global scientific advancements.

2.5.3 The Disconnect Between Policy Decisions and Real Research Needs

One of the biggest failures of India's research ecosystem is the disconnect between policymakers and researchers. Government bodies that control research funding often lack direct engagement with scientists and independent scholars, leading to misaligned priorities, inefficient spending, and lost opportunities.

1. Research Funding Priorities Are Often Driven by Politics, Not Science

- The Indian government frequently announces new research initiatives without consulting actual scientists or industry experts.
- Many funding decisions are influenced by political agendas, resulting in misallocation of resources to low-impact or irrelevant projects.
- In contrast, countries like Germany and the US involve scientists in research policymaking, ensuring that funding aligns with the needs of the research community.

2. Lack of Representation of Independent Researchers in Policy Discussions

- Indian research policies are primarily shaped by institutional committees, with little to no input from independent researchers.

- Unlike in the US and EU, where researchers are regularly consulted for policy design and research roadmaps, India's policy-making process excludes the very people it is meant to support.

3. Wastage of Research Funds Due to Poor Allocation

- India spends billions on institutional R&D programs, yet many labs remain underutilized, and projects lack direction.
- A significant percentage of government research grants remain unspent due to bureaucratic inefficiencies, while independent researchers struggle to get even basic funding for their work.

Conclusion

The neglect of independent researchers in India is not just a policy failure—it is a national crisis. While other countries fund, nurture, and empower their independent scholars, India actively discourages them through bureaucratic barriers, institutional bias, and political misalignment.

The result?

- Massive brain drain, as talented researchers leave for better opportunities abroad.
- Lost innovation, as bureaucratic delays kill high-impact projects.
- Misdirected funding, as policymakers fail to align research spending with actual needs.

If India wants to become a global leader in science and innovation, it must stop driving away its best minds and restructure its research ecosystem to support independent researchers with direct funding, fewer bureaucratic hurdles, and meaningful policy representation.

Without these changes, India will remain a nation of lost potential, watching from the sidelines as other countries take the lead in scientific discovery and technological advancement.

2.6 The Way Forward: Breaking the Institutional Barrier for Independent Researchers

India's rigid, institution-dominated research ecosystem has systematically sidelined independent researchers, stifling innovation, driving away talent, and preventing high-impact discoveries from materializing. If India is serious about becoming a global leader in science, technology, and innovation, it must break institutional barriers and create a research ecosystem that empowers both institutional and independent scholars alike.

This requires a radical policy shift that includes direct funding for independent researchers, streamlined research fellowships, and stronger industry collaborations. Without these reforms, India will continue to lose its brightest minds to countries that actively support independent research.

2.6.1 Proposed Policy Reforms to Make Funding Accessible for Individuals

To foster an inclusive research ecosystem, India must redefine its funding policies so that independent researchers receive the same opportunities as institutional scholars. This can be achieved through:

1. Introducing Independent Research Grants

- Government funding agencies like DST, SERB, CSIR, and ICMR should introduce dedicated funding schemes for independent researchers.
- These grants should allow individuals without university

affiliation to apply based on merit, research feasibility, and impact potential rather than institutional backing.

- A transparent review system should be implemented to ensure fair evaluation of research proposals from independent scholars.

2. Reducing Bureaucratic Barriers in Grant Approvals

- The multi-layered approval process for research grants should be streamlined to cut unnecessary red tape.
- A single-window grant application process should be introduced to ensure faster processing and disbursement of funds.
- Government committees evaluating research proposals should include independent researchers, scientists, and industry experts, rather than just institutional administrators.

3. Establishing an Independent Research Fund

- India should create a National Independent Research Fund (NIRF), modeled after the US National Science Foundation (NSF) or Germany's Deutsche Forschungsgemeinschaft (DFG).
- The fund should be autonomous from institutional biases and open to applications from individual researchers, startups, and non-affiliated scholars.
- This initiative should focus on high-impact research areas, including AI, biotechnology, renewable energy, space exploration, and social sciences.

4. Providing Research Tax Benefits and Incentives

- Independent researchers should receive tax deductions on

expenses related to research, including equipment purchases, lab access, journal subscriptions, and patent filings.

- Self-funded research projects should qualify for government subsidies, similar to how R&D tax credits are provided to corporations.
- Researchers who develop patentable innovations or tech startups based on their independent research should receive financial incentives to commercialize their work.

By removing institutional gatekeeping from research funding, India can unlock the full potential of its independent thinkers and ensure that groundbreaking research is no longer restricted to a select group of institutions.

2.6.2 Creating Direct Research Fellowships for Non-Affiliated Researchers

Currently, almost all Indian fellowships—such as UGC NET-JRF, CSIR Fellowships, and ICMR Fellowships—are exclusively available to institutional researchers. This exclusionary practice must end if India wants to encourage a thriving independent research culture.

1. Establishing National Fellowships for Independent Scholars

- A National Independent Research Fellowship (NIRF) should be created for self-funded researchers, allowing them to pursue full-time research without the need for university affiliation.
- Fellowships should be merit-based and focus on supporting researchers in underfunded fields such as basic sciences, humanities, social sciences, and disruptive technologies.

2. Offering Infrastructure Access to Independent Researchers

- Independent researchers should be given access to

government research labs, supercomputing facilities, and scientific databases at subsidized costs or for free.

- A Research Access Card should be introduced, allowing independent scholars to use institutional research resources without formal enrollment in an institution.
- Open-access research networks should be established, modeled after Germany's Max Planck Society, where researchers can apply for short-term lab access or project-based collaborations.

3. Encouraging Collaboration Between Institutional and Independent Researchers

- Indian research institutions should be mandated to allocate a portion of their funding and facilities for collaborations with independent scholars.
- Universities should introduce Visiting Researcher Programs, where independent researchers can work alongside institutional faculty without requiring permanent affiliation.
- The government should facilitate interdisciplinary research hubs, where independent scholars, industry professionals, and academic researchers can collaborate on high-impact projects.

By creating a strong support structure for independent researchers, India can increase its global research contributions and encourage a culture of open innovation.

2.6.3 Encouraging Industry Partnerships to Fund Independent Research

One of the biggest reasons why India lags behind in independent research is the lack of private-sector investment in non-institutional R&D. While US and European companies actively fund independent innovators, Indian corporations remain disengaged from research sponsorships.

To bridge this gap, strong industry-academia collaborations must be established that allow companies to directly invest in independent research.

1. Creating a Corporate Research Sponsorship Program

- The government should introduce tax incentives for private companies that fund independent researchers, similar to R&D tax credit programs in the US and Germany.
- Corporations should be encouraged to set up independent research sponsorship programs, providing grants and mentorship to self-funded researchers working on industry-relevant projects.

2. Establishing Public-Private Research Incubators

- The government should create Public-Private Research Incubation Centers, where independent researchers can receive corporate sponsorships, mentorship, and infrastructure support.
- These incubators should focus on high-growth sectors such as artificial intelligence, blockchain, clean energy, and medical research.
- Leading Indian corporations (such as Tata, Infosys, and Reliance) should be encouraged to invest in these open research ecosystems.

3. Opening Crowdfunding and Alternative Research Funding Models

- Independent researchers should have access to alternative funding sources, including:

- Crowdfunding platforms (similar to Experiment.com, where individuals can raise funds from the public).
- Angel investment networks that support scientific research startups.
- Venture capital funding for high-risk, high-reward research.
- The government should legalize and regulate research crowdfunding, ensuring that independent researchers can receive public donations without bureaucratic interference.

By incentivizing industry involvement in independent research, India can create a self-sustaining ecosystem where researchers do not have to rely solely on government grants.

Conclusion

Breaking the institutional monopoly on research funding is not just about equity—it is about unlocking India's true intellectual potential. Independent researchers have long been ignored, underfunded, and systematically excluded from India's research ecosystem, leading to brain drain, lost innovations, and scientific stagnation.

To reverse this trend, India must:

- Create direct research grants and fellowships for independent scholars.
- Simplify the funding application process to remove unnecessary bureaucracy.
- Provide tax benefits and infrastructure access to self-funded researchers.
- Encourage industry partnerships, corporate sponsorships, and alternative research funding models.

Without these reforms, India will remain a follower rather than a leader in global scientific advancements. But with the right policies,

India can transform itself into an innovation powerhouse, where independent researchers play a critical role in shaping the future of science and technology.

3. The Research Infrastructure Crisis

Scientific progress thrives on access to knowledge, laboratories, and data—yet, in India, independent researchers face severe restrictions in all these areas. Unlike their institutional counterparts, who benefit from fully funded labs, extensive research databases, and exclusive journal subscriptions, independent researchers are left struggling to access even the most basic scientific resources.

This systemic denial of infrastructure is not accidental; it is a result of institutional monopoly, bureaucratic policies, and outdated access restrictions. Government labs, university research centers, and national knowledge repositories are reserved for a privileged few, leaving independent scholars without the tools they need to conduct meaningful research. The situation is further worsened by expensive journal paywalls, restricted access to crucial data, and the lack of open research platforms, making it nearly impossible for self-funded researchers to keep up with advancements in their fields.

This chapter critically examines how India's infrastructure crisis is suffocating independent research, exploring the barriers to lab access, institutional control over scientific resources, and restrictive policies that prevent access to crucial data. Unless these systemic flaws are addressed, India will continue to hinder its own intellectual potential, allowing only a small fraction of researchers to contribute to scientific discovery while the rest remain locked out of the system.

3.1 The Lockout: Why Independent Researchers Lack Access to Labs, Journals, and Libraries

Access to cutting-edge research infrastructure—including well-equipped laboratories, scientific journals, and academic libraries—is fundamental to conducting high-quality research. However, in India, independent researchers are systematically locked out of these essential resources. The lack of access is not merely an oversight; it is a deliberate consequence of institutional policies, financial barriers, and government inaction.

Unlike researchers affiliated with universities or government-backed labs, independent scholars must navigate an expensive, restrictive, and exclusionary system that leaves them without the tools needed to produce impactful research. From overpriced journal subscriptions to exclusive institutional memberships and outdated access policies, these barriers effectively prevent independent researchers from contributing to the country's knowledge economy.

3.1.1 The Cost Barrier: How Expensive Journal Subscriptions and Lab Access Prevent Independent Research

One of the biggest obstacles independent researchers face is the prohibitive cost of accessing scientific literature and laboratory facilities. Unlike institutional researchers, who receive free access to libraries, research labs, and journal subscriptions through government funding, independent scholars must pay exorbitant fees for the same resources.

1. Journal Subscription Paywalls: The Cost of Knowledge

- Most prestigious scientific journals, including Nature, Science, IEEE, Elsevier, and Springer, require paid subscriptions for

access to research papers.
- Subscription costs range from ₹1,00,000 to ₹5,00,000 per year, making it financially impossible for independent researchers to afford them.
- Unlike the US, Germany, or the UK, where universities often provide discounted or subsidized journal access to independent scholars, Indian institutions restrict access to their own members only.
- As a result, independent researchers are forced to rely on outdated or incomplete data, reducing the quality of their work.

2. Expensive Lab Access and Research Facilities

- Scientific research requires access to high-end laboratories, testing centers, and computing resources, most of which are controlled by government or university institutions.
- Even when access is granted, usage fees are prohibitively high.
- For example, private biotech labs charge anywhere between ₹10,000 to ₹50,000 per session for independent research, making long-term projects unsustainable.
- Advanced computing facilities for AI, quantum research, and simulations often charge per usage hour, adding further financial strain.

3. No Government Subsidy for Independent Researchers

- Unlike corporate R&D or university research projects, independent researchers do not receive any financial aid or subsidies from the government.
- In countries like Germany and the US, independent researchers can apply for subsidized access to public research infrastructure, but India offers no such program.

The direct consequence of this financial barrier is that only well-funded institutions and corporate-backed researchers can afford to pursue high-quality research, while independent scholars are forced to abandon their studies or rely on outdated methodologies.

3.1.2 Institutional Exclusivity: How Universities and Research Institutes Restrict Access to Their Resources

India's publicly funded universities and research institutions, which should ideally serve as knowledge-sharing hubs, operate under a closed-access model that favors institutional researchers over independent scholars. These institutions control most of the country's research resources, yet they actively deny access to non-affiliated researchers, creating an academic monopoly that stifles intellectual diversity.

1. Universities Restricting Library and Journal Access

- Most IITs, IISc, NITs, and Central Universities receive government grants to purchase journal subscriptions and maintain research databases.
- Despite being publicly funded, these institutions restrict access to enrolled students and faculty members only.
- Independent researchers cannot use university libraries or digital repositories without affiliation, even though these resources are funded by taxpayer money.
- Countries like Germany and the Netherlands have introduced publicly accessible digital libraries, but India has no such initiative.

2. Research Labs Denying External Access

- Most government-funded research labs refuse to collaborate with independent researchers unless they are part of an institutional project.

- While some facilities allow temporary access through MoUs with private firms, individual researchers cannot apply for lab usage.
- Even when access is possible, bureaucratic approvals can take months, delaying critical research progress.

3. Faculty Bias Against Independent Researchers

- Many university faculty members prefer to work with affiliated students rather than independent researchers, fearing that collaborations outside their institution may not be recognized by funding bodies.
- In contrast, countries like the US and Canada encourage cross-institutional and independent collaborations, allowing researchers to work across academic and private research organizations.

The exclusionary nature of Indian universities and research institutions has made it nearly impossible for independent researchers to participate in serious scientific work, forcing them to either abandon their research or seek foreign collaborations.

3.1.3 Lack of Open Access Platforms: Why India Has Not Adopted Open Research Models Like the US and Europe

Globally, many governments have recognized the need for open-access platforms to ensure that research findings are available to all, including independent researchers, startups, and policymakers. However, India has failed to implement any significant open-access initiative, further isolating independent scholars.

1. The Open-Access Revolution in Developed Countries

- The United States, under initiatives like the Public Access

Policy, mandates that government-funded research be made freely accessible online.

- The European Union has launched Plan S, which requires all publicly funded research to be published in open-access journals rather than behind paywalls.
- Germany's Max Planck Society has set up open-access repositories, allowing researchers worldwide to use their data.

2. India's Failure to Create Open Research Platforms

- Despite receiving billions in research funding, India does not require publicly funded institutions to share their research openly.
- Most research papers produced under government grants are still locked behind paywalls, benefiting private journal publishers rather than the public.
- The absence of a national research repository means that even Indian researchers must pay to access work funded by their own government.

3. The Need for a National Open Research Platform

- India must create an open-access database where all government-funded research, patents, and academic papers are freely available.
- Public research institutions must be mandated to upload their studies to a centralized repository, similar to arXiv.org in the US.
- Private firms benefiting from government-funded research should also be required to contribute findings to the open-access platform.

By failing to establish open-access research platforms, India is denying its independent researchers access to critical knowledge and scientific breakthroughs, limiting their ability to innovate and compete globally.

Conclusion

Independent researchers in India are not just struggling—they are actively being locked out of the scientific ecosystem. Unlike institutional researchers, who have free access to labs, libraries, and journals, independent scholars are forced to pay exorbitant fees, denied entry to research centers, and excluded from knowledge-sharing networks.

This crisis has led to:

- The exclusion of independent voices from India's scientific community.
- A growing dependence on foreign resources, as researchers turn to international collaborations.
- A widening gap between institutional and independent research capabilities.

To fix this, India must adopt global best practices by:

- Mandating open-access publishing for publicly funded research.
- Allowing independent researchers access to university libraries and government-funded labs.
- Creating subsidized infrastructure access programs for non-affiliated scholars.

If these reforms are not implemented, India will continue to isolate its brightest minds, ensuring that groundbreaking research happens elsewhere while Indian innovation stagnates behind institutional walls.

3.2 Institutional Monopoly on Scientific Resources

Scientific resources—whether in the form of laboratories, research equipment, academic journals, or funding—should ideally be accessible to all researchers contributing to knowledge and innovation. However, in India, these resources are monopolized by institutions, leaving independent researchers without access to essential tools needed for high-quality scientific work.

This institutional monopoly is not just a logistical issue; it is a structural barrier that forces independent scholars to either give up on research or seek opportunities abroad. The restrictions on lab access, the financial burden of journal subscriptions, and the funding bias toward institutional projects ensure that only a privileged few within formal academic structures can engage in meaningful research.

3.2.1 Restricted Lab Access: How Government and University Labs Deny Independent Researchers

In developed research ecosystems, national and university laboratories are often open to independent researchers, startups, and cross-institutional collaborations. However, in India, publicly funded research facilities are largely inaccessible to non-affiliated scholars.

1. The Illusion of Public Research Facilities

- Many of India's top research institutions (IITs, IISc, DRDO, ISRO, CSIR labs, etc.) receive government funding to develop cutting-edge labs.
- Despite being publicly funded, these labs are restricted to enrolled students, faculty members, or government-affiliated projects.
- Independent researchers cannot access these labs unless they have an institutional collaboration or special approval, both of which involve bureaucratic hurdles.

2. The Bureaucratic Nightmare of Gaining Lab Access

Even when independent researchers try to collaborate with an institution to gain lab access, they must go through lengthy approval processes.

- Researchers often face months of administrative delays, requiring:
- Endorsements from university faculty.
- Clearance from multiple committees.
- A formal institutional tie-up, which many universities refuse.

3. The Impact on Scientific Progress

- The lack of lab access forces many independent researchers to abandon experimental work, limiting them to theoretical research or computational studies.
- In countries like Germany, the US, and China, national laboratories allow external researchers to book time slots for experiments—a model India has yet to implement.
- Without accessible laboratory infrastructure, independent researchers in India are left to rely on personal resources, which are insufficient for advanced scientific exploration.

3.2.2 Paywalls and Membership Barriers: The Role of Paid Journal Subscriptions in Limiting Research

Access to scientific literature and academic publications is essential for researchers to build on existing knowledge. However, most high-impact research papers are locked behind paywalls, making them unaffordable for independent scholars.

1. The Cost of Scientific Knowledge

- Journals like Nature, Science, IEEE, and Elsevier charge hefty subscription fees, often running into lakhs of rupees per year.

- University researchers receive free access through their institutions, but independent scholars must pay out of pocket.
- A single research paper can cost between ₹3,000 and ₹8,000, making it financially unsustainable to access multiple papers.

2. Why India Lags in Open-Access Initiatives

- Unlike countries such as Germany, the Netherlands, and the UK, where governments have implemented mandatory open-access policies, India has no such national initiative.
- Research funded by Indian taxpayers remains behind paywalls, benefiting private publishers rather than the scientific community.

3. The Digital Divide: Limited Library Access for Non-Affiliated Scholars

- University libraries maintain subscriptions to high-impact journals, but only students and faculty can access them.
- Independent researchers cannot use digital archives like Scopus, Web of Science, or IEEE Xplore unless they have an institutional login.
- In contrast, countries like the US and Germany have national open-access research repositories, allowing all scholars to download publicly funded research for free.
- The exclusivity of scientific literature further isolates independent researchers, preventing them from staying updated with the latest developments in their fields.

3.2.3 Funding Bias in Infrastructure: Why Major Research Facilities Prioritize Institutional Projects Over Individual Studies

In India, scientific funding and infrastructure development are concentrated in a few elite institutions, creating a research hierarchy that favors large organizations while neglecting individual innovators.

1. Centralized Research Funding: The Institutional Advantage

- Government research grants, infrastructure development funds, and R&D budgets are mostly allocated to IITs, IISc, NITs, and CSIR labs.
- These institutions receive billions in funding every year, while independent researchers struggle to secure even basic financial support.
- Unlike in the US, where the National Science Foundation (NSF) funds independent scholars, India has no equivalent structure for non-affiliated researchers.

2. Limited Industry Collaboration for Non-Institutional Researchers

- Private companies often collaborate with well-known institutions for R&D projects, overlooking independent researchers.
- Indian startups and tech firms rarely invest in independent research, unlike in the US and Europe, where companies frequently fund individual innovators.

3. The Disconnect Between Government Funding and Research Needs

- Government-funded research facilities often remain underutilized because of restrictive access policies.
- Large institutions monopolize resources, while independent researchers struggle to conduct meaningful experiments due to lack of funding.

- Countries like China and Germany have introduced shared research models, allowing individual scholars to access government labs—a concept that India has yet to embrace.

The funding imbalance in research infrastructure ensures that only a select group of institutions benefit from national resources, while independent researchers are left without essential facilities.

Conclusion

India's institutional monopoly on scientific resources has created a two-tiered research ecosystem, where institutional scholars enjoy privileged access to labs, journals, and funding, while independent researchers face insurmountable barriers.

The consequences of this exclusivity are severe:

- Independent researchers are locked out of essential scientific tools and knowledge.
- India loses innovative thinkers to foreign institutions with better access policies.
- Publicly funded research remains inaccessible to the very scholars it should support.

To break this monopoly, India must:

- Allow independent researchers to use government-funded labs and libraries.
- Introduce national open-access policies for publicly funded research.
- Create funding mechanisms specifically for non-affiliated scholars.

Without these changes, India will continue to stifle its own research potential, ensuring that scientific breakthroughs happen elsewhere while Indian scholars remain restricted by institutional gatekeeping.

3.3 Outdated Policies That Restrict Access to Crucial Data

Access to accurate, up-to-date research data is the foundation of scientific progress. In an era where information fuels innovation, countries across the world are investing in open-access platforms, digital archives, and data-sharing initiatives to ensure that researchers—both institutional and independent—can freely access national and global research findings.

However, India remains stuck in an outdated, bureaucratic model where research data is heavily restricted, hidden behind paywalls, or simply inaccessible. The lack of public research archives, excessive bureaucracy, and the absence of a national open-access policy has made it extremely difficult for independent researchers to access crucial scientific data. This restrictive approach is pushing India further behind in global research rankings, as many of its brightest minds struggle to access even publicly funded research outputs.

3.3.1 Bureaucratic Roadblocks: Why National Research Data is Not Freely Available

In many developed nations, research data—particularly government-funded studies and scientific findings—is made freely available to all researchers, policymakers, and innovators. However, in India, a complex web of bureaucratic regulations and institutional controls ensures that even publicly funded research remains inaccessible to independent researchers.

1. Government-Imposed Access Restrictions

- Research funded by agencies like CSIR, ICMR, ISRO, DRDO, DST, and DBT is often classified or restricted, even if it has no national security implications.

- Independent researchers are forced to go through lengthy approval processes to gain access to datasets, often facing denials with no explanation.
- In contrast, countries like the US and UK have mandatory transparency policies that require all publicly funded research to be accessible to the public.

2. Excessive Administrative Procedures and Delays

- Even when researchers apply for official access to research data, the process can take months or even years due to slow-moving bureaucratic systems.
- Some researchers are asked to submit multiple requests through different departments, with no single-window mechanism to streamline access.
- By the time access is granted, the data may already be outdated, making it useless for cutting-edge research.

3. Institutional Monopoly Over Government-Funded Data

- Universities and research institutions that receive government grants control access to research findings, refusing to share them with external researchers.
- Even when independent scholars seek collaborations, institutions demand formal affiliations, making it impossible for truly independent research to be conducted.
- In many cases, even university faculty members struggle to access government research data because of complex bureaucratic approvals.

This bureaucratic control over scientific data is delaying discoveries, slowing innovation, and discouraging independent research in India.

3.3.2 No Public Research Archives: The Absence of an Open-Access Database for Indian Research Papers

One of the biggest disadvantages Indian researchers face is the lack of a centralized, open-access database for government-funded research. In most developed nations, governments have created public repositories where all state-funded research is freely available to scientists, students, policymakers, and entrepreneurs. However, India has failed to implement such a system, keeping research data locked within institutions and private journal paywalls.

1. No National Open-Access Repository for Research Papers

- Countries like the US (PubMed, arXiv, and NIH database) and the EU (CORDIS, OpenAIRE) have built large-scale public databases for government-funded research.
- In contrast, India has no centralized digital platform where researchers can access government-funded studies, patents, and datasets.
- As a result, independent researchers are forced to rely on outdated books, expensive journal subscriptions, or fragmented private databases.

2. The Problem with Privately Owned Journal Paywalls

- Much of India's scientific research is published in foreign journals like Elsevier, Springer, and Wiley, where access fees can range from ₹3,000 to ₹8,000 per paper.
- Since India does not mandate open-access publishing for taxpayer-funded research, many publicly funded studies end up behind expensive paywalls.
- In contrast, the European Union's Plan S requires all research funded by public grants to be freely available to the public.

3. No Open-Data Initiative for Public Research Institutions

- Government agencies like CSIR, ICMR, and DRDO have thousands of research publications and technical reports that remain hidden from the public.
- Instead of creating a free and transparent database, India suffers from fragmented research archives, making it almost impossible for independent researchers to find useful data.

The lack of an open-access national research archive has severely crippled India's ability to foster innovation, as researchers are denied access to the very knowledge they need to advance their fields.

3.3.3 Global Best Practices: How Countries Like the US and Germany Allow Open Access to Research Findings

While India continues to restrict access to research data, other nations have embraced open-access policies, ensuring that publicly funded research is available to all researchers, entrepreneurs, and innovators.

1. The United States: Open-Access Research Mandates

- The National Institutes of Health (NIH) requires all publicly funded research papers to be freely available through PubMed Central.
- The National Science Foundation (NSF) and NASA operate open-access portals where scientific data, research findings, and technical reports are freely accessible.
- The US government has introduced policies preventing taxpayer-funded research from being locked behind private journal paywalls.

2. The European Union: Plan S and OpenAIRE

- The European Commission launched Plan S, a policy requiring all publicly funded research to be published in open-access journals.
- The OpenAIRE platform provides a free and centralized research database, allowing independent researchers to download government-funded studies without restrictions.
- Unlike India, where institutions hoard publicly funded research, Europe ensures knowledge is accessible to everyone.

3. Germany: The Max Planck Society and Open Research Initiatives

- The Max Planck Society runs a global open-access movement, making research available to independent scholars worldwide.
- Germany's Deutsche Forschungsgemeinschaft (DFG) funds open-access journals to ensure that all state-funded studies remain free for public use.
- Unlike India, Germany does not allow universities to block access to taxpayer-funded research.

India has no excuse for not implementing similar initiatives. If other countries have successfully created open-access databases, there is no reason why Indian researchers should continue to struggle with data restrictions and journal paywalls.

Conclusion

India's failure to provide open access to scientific research is one of the biggest obstacles preventing independent scholars from conducting meaningful research. Publicly funded studies, government research data, and technical findings are all locked behind paywalls and bureaucratic red tape, preventing researchers from advancing knowledge, developing solutions, or competing globally.

If India wants to unlock its research potential, it must:

- Create a national open-access research database where all publicly funded studies are freely available.
- Mandate that research funded by taxpayer money must be published in open-access journals, not locked behind paywalls.
- Simplify bureaucratic processes so that independent researchers can access government-funded datasets without excessive approvals.

Without these reforms, India will continue to fall behind in global scientific progress, while countries like the US, Germany, and China lead the way in innovation, research, and knowledge-sharing.

4. Industry's Neglect: Why Indian Corporate Avoid R&D

India has long been recognized as a global hub for manufacturing, IT services, and pharmaceuticals, but when it comes to original research and innovation, the country lags far behind. Despite government campaigns like "Make in India", the private sector continues to prioritize production over research, relying heavily on imported technology, foreign patents, and outsourced expertise. Unlike global counterparts in the US, Germany, China, and Japan, where industries actively invest in R&D, Indian corporations see research as an unnecessary cost rather than a long-term investment.

This reluctance to fund in-house research has created a dangerous cycle of technological dependence, where Indian companies prefer importing technology rather than developing their own. The weak collaboration between industry and academia further widens the innovation gap, leaving India struggling to compete in high-tech industries like semiconductors, AI, and biotechnology.

This chapter explores the systemic failures in India's corporate R&D ecosystem, examining why industries avoid investing in research, how they rely on foreign technology, and the missed opportunities that have cost India its place in global innovation. Without a shift in mindset—where research is seen as a critical driver of economic growth rather than an unnecessary expense—India risks remaining a follower rather than a leader in the global innovation race.

4.1 The Myth of "Make in India" Without "Research in India"

The "Make in India" campaign was launched with the promise of transforming India into a global manufacturing powerhouse. The initiative aimed to boost industrial production, attract foreign investment, and create jobs. However, while "Make in India" has succeeded in expanding manufacturing and assembly lines, it has done little to encourage indigenous research and technological innovation.

For India to truly become self-reliant in manufacturing, it must transition from merely assembling products with imported technology to developing homegrown innovations. Unfortunately, India's industrial policies continue to prioritize mass production over deep R&D, leaving the country dependent on foreign patents, licensing agreements, and technological imports. This section explores why India's focus on manufacturing without research is a flawed strategy, how government policies contribute to this imbalance, and the long-term consequences of weak domestic innovation.

4.1.1 Manufacturing vs. Innovation: Why India Focuses on Production but Neglects Original Research

India has successfully positioned itself as a global hub for manufacturing, especially in automobiles, pharmaceuticals, electronics, and textiles. However, despite strong industrial growth, the country has failed to create a research-driven industrial ecosystem.

1. The Preference for Low-Cost Manufacturing

- Many Indian industries operate on thin profit margins, making them reluctant to invest in high-risk, long-term research projects.
- Rather than focusing on breakthrough technologies, most companies opt for cost-cutting measures and process improvements to remain competitive.

- Countries like South Korea and Japan have invested heavily in technology-driven industrial expansion, while India has focused on low-cost, high-volume production.

2. The "Assembly Economy" Model

- India has become a major center for assembling smartphones, automobiles, and consumer electronics, but very few of these products use Indian-made components or patented technology.
- While companies like Apple, Samsung, and Tesla manufacture in India, they import key technologies, such as semiconductors, battery cells, and advanced sensors.
- In contrast, countries like China and Taiwan have established indigenous supply chains to support research-led manufacturing.

3. The Gap in Research Spending

- India's Gross Expenditure on Research and Development (GERD) remains below 1% of GDP, compared to: 2.8% in the US, 2.4% in China, 4.5% in South Korea
- Without a strong research foundation, Indian industries remain dependent on foreign technology and struggle to develop competitive intellectual property (IP).

While mass production helps in short-term economic growth, the lack of research investment prevents India from becoming a true innovation leader.

4.1.2 Government Policies Favoring Assembly-Line Production Over Indigenous R&D

Government policies over the past decades have favored large-scale manufacturing and industrial expansion but have largely ignored research incentives for private industries.

1. Focus on Foreign Direct Investment (FDI) Instead of Local Innovation

- India's manufacturing boom has been driven by foreign companies setting up assembly units, but very few of them invest in R&D within India.
- Policies encourage FDI for manufacturing but lack strong incentives for technology transfer and research partnerships.
- For Example: Apple's manufacturing partners (Foxconn, Pegatron) have opened large factories in India but continue to source high-tech components from China and Taiwan.

2. Weak Research Incentives for Indian Industries

- Indian industries receive tax breaks for production and exports, but there are limited incentives for in-house R&D spending.
- The R&D tax credit in India was reduced from 200% to 150% in 2016 and further cut to 100% in 2020, making it less attractive for companies to invest in research.
- In contrast, China provides tax deductions of up to 175% for companies investing in R&D, driving rapid industrial innovation.

3. Limited Government-Industry Research Collaboration

- Unlike countries like Germany and the US, where governments actively fund private-sector research, India's collaboration between industry and government-funded labs remains weak.

- Public sector research organizations like CSIR, DRDO, and ISRO develop technologies, but very few are commercialized by Indian industries.
- The disconnect between academic research and industrial needs means that many innovations remain unused or are commercialized abroad.

Government policies must shift from promoting only manufacturing to incentivizing industrial R&D, enabling Indian companies to develop cutting-edge technologies rather than just assembling imported ones.

4.1.3 The Consequences of Weak Domestic Innovation on Long-Term Economic Growth

India's failure to develop a strong research-driven industrial sector has long-term economic consequences. The absence of indigenous innovation has led to:

1. Dependence on Foreign Technology

- Indian industries rely heavily on imported intellectual property (IP), software, and hardware from the US, Europe, China, and Japan.
- This dependence increases trade deficits, as companies must pay billions in licensing fees to foreign firms.
- For Example: India imports over 90% of its semiconductor requirements, making it vulnerable to global supply chain disruptions.

2. Missed Opportunities in Emerging Technologies

- While India has strong IT services and pharmaceutical industries, it has failed to lead in:
- Semiconductors (Taiwan and South Korea dominate).

- AI and Machine Learning (The US and China are ahead).
- Electric Vehicles (China has built a self-sufficient EV supply chain).

Without in-house research and product development, Indian companies struggle to compete in the global high-tech market.

3. Limited High-Skill Job Creation

- Research-driven industries create high-paying jobs for scientists, engineers, and developers.
- By focusing on manufacturing and assembly rather than R&D, India is not producing enough high-value jobs.
- For Example: South Korea's investment in R&D led to the rise of Samsung, Hyundai, and LG, creating millions of high-tech jobs.

4. Vulnerability to Global Market Shifts

- Countries that fail to invest in innovation become dependent on others for technological progress.
- If global trade policies shift (e.g., restrictions on technology imports), India's industries will struggle to adapt.
- Without a self-sustaining R&D ecosystem, India will always be a step behind in the race for technological dominance.

Conclusion

The "Make in India" initiative has expanded India's manufacturing base, but it has failed to create a culture of research and innovation. The focus on assembly-line production, reliance on foreign technology, and lack of industrial R&D investment has made India a follower rather than a leader in global innovation.

- To shift from "Make in India" to "Innovate in India," the

country must:

- Increase R&D tax benefits for private industries.
- Encourage technology transfer and research partnerships in FDI policies.
- Create stronger industry-academia collaboration for commercializing innovations.
- Invest in next-generation research fields like AI, semiconductors, and clean energy.

If these changes are not implemented, India will continue to assemble products for global companies while failing to develop its own world-class innovations. The future of India's industrial growth depends on whether it can move beyond manufacturing and embrace research as a fundamental pillar of economic progress.

4.2 How Companies Import Technology Instead of Funding Research

India's corporate sector has built a strong presence in manufacturing, IT services, and pharmaceuticals, yet when it comes to original research and product development, the country remains heavily dependent on foreign technology. Instead of investing in long-term R&D, Indian companies prefer to license patents, import solutions, or collaborate with foreign firms for technology transfer.

While this approach provides short-term cost savings and faster production cycles, it creates a long-term dependency on external innovation, preventing India from becoming a global leader in high-tech industries. This section explores why Indian companies avoid in-house R&D, the consequences of over-reliance on foreign technology, and the cost-benefit trade-offs involved in this approach.

4.2.1 Over-Reliance on Foreign Patents and Imported Technology Solutions

India's industrial growth has largely been driven by technology imports rather than indigenous innovation. While companies across the globe invest in patents, proprietary technologies, and R&D-driven product development, many Indian firms take a shortcut by licensing existing technology from foreign companies.

1. Technology Import Trends in Key Sectors

- **Electronics and Semiconductor Industry:** India imports over 90% of its semiconductors from China, Taiwan, and the US. No major Indian company has successfully developed an indigenous semiconductor chip, despite the government's push for local manufacturing.
- **Pharmaceutical Industry:** While India is a leading producer of generic drugs, it imports active pharmaceutical ingredients (APIs) from China. Very few Indian firms invest in original drug discovery, preferring to replicate existing formulas rather than create new ones.
- **Automobile and EV Sector:** Indian carmakers rely on foreign patents for battery technology, electric motors, and advanced electronics. Even companies like Tata Motors and Mahindra depend on global suppliers for high-end vehicle components.
- **Defense and Aerospace Industry:** India imports over 60% of its defense equipment, making it one of the largest arms importers in the world. Despite having ISRO and DRDO, private defense R&D remains underdeveloped, with most components sourced from foreign suppliers.

2. The Problem with a Technology-Import Economy

High Royalty and Licensing Costs: Indian companies pay billions in licensing fees to access foreign technology. **For Example:** Automobile firms pay heavy royalties to global automakers for using patented designs and engines.

Lack of Control Over Critical Technologies: Industries relying on imported technology are vulnerable to global supply chain disruptions. **For Example**: The COVID-19 pandemic led to shortages of APIs, severely affecting India's pharmaceutical sector.

Missed Opportunities in Global Patents and Intellectual Property: While China, the US, and South Korea dominate global patent filings, India's share remains extremely low. Fewer patents mean that India remains a consumer of technology rather than a creator of new innovations.

4.2.2 Why Indian Companies Prefer Licensing Foreign Technology Rather Than Developing Their Own

Rather than funding their own research labs, Indian companies prefer to license technology from established global players. Several factors contribute to this preference:

1. The "Low-Risk, High-Reward" Mindset

- Developing new technologies requires high-risk investment with no guarantee of success.
- Licensing existing technology allows companies to produce goods quickly without spending years in R&D.
- This model reduces financial risk but prevents companies from owning proprietary technology.

2. The Short-Term Profitability Model

- Indian businesses prioritize quarterly financial performance over long-term R&D investments.
- Unlike countries like Japan or Germany, where firms invest decades in technological advancements, Indian corporations focus on immediate profitability.
- Example: The automobile industry prefers foreign battery technology for EVs rather than developing indigenous

alternatives.

3. The Absence of Strong R&D Ecosystems

- Unlike China, the US, and Germany, India lacks dedicated industrial research parks and technology incubators.
- Private companies struggle to find skilled R&D talent, as most top researchers move abroad for better opportunities.
- Government R&D institutions like CSIR, DRDO, and ISRO are largely disconnected from corporate research needs, leading to limited industry-academia collaboration.

4. The Comfort of Technology Transfers from Foreign Companies

- Indian businesses often secure technology transfers from multinational corporations instead of investing in their own R&D.
- Example: India's defense sector heavily relies on Russian, American, and Israeli technology transfers rather than developing indigenous systems.
- This model creates dependency and prevents long-term self-sufficiency.

While licensing technology provides quick access to advanced solutions, it fails to build India's industrial research capabilities, leaving the country permanently reliant on foreign innovation.

4.2.3 The Cost-Benefit Analysis: Short-Term Gains vs. Long-Term Dependence

While importing technology helps companies avoid R&D risks, it comes at the cost of long-term innovation and industrial self-sufficiency.

1. Short-Term Gains of Licensing and Importing Technology

- **Lower upfront investment:** Companies can start production faster without spending on R&D.
- **Faster market entry:** Imported technologies allow businesses to launch products quicker.
- **Lower failure rates:** Licensing minimizes the risk of failed research projects.

2. Long-Term Risks of Avoiding R&D

- **Perpetual foreign dependence:** Indian industries remain vulnerable to supply chain disruptions.
- **Loss of economic value:** Indian companies pay billions in royalties and technology fees to foreign firms.
- **Weaker global competitiveness:** Without strong R&D, Indian firms struggle to develop high-tech exports.

3. Lessons from Countries That Prioritized R&D

- **South Korea:** Instead of relying on foreign technology, Samsung and LG invested heavily in R&D, making them leaders in semiconductors, AI, and display technologies.
- **China:** The government mandated a "Made in China 2025" strategy, investing billions in local research parks and technology hubs to reduce foreign dependence.
- **Germany:** Strong industry-academia collaboration ensures that German industries lead in automotive and engineering innovations.

India can no longer afford to rely solely on technology imports. To become a global R&D powerhouse, Indian companies must:

- Invest in indigenous research and patent development.
- Build long-term R&D ecosystems within industries.

- Strengthen collaboration between government labs and private businesses.
- Encourage risk-taking in industrial research.

Conclusion

India's corporate sector cannot innovate without investing in research. The over-reliance on foreign patents, licensing agreements, and imported technology has stunted India's ability to create its own cutting-edge industries.

If Indian businesses continue to avoid in-house research, the country will:

- Remain dependent on foreign suppliers.
- Struggle to compete in high-tech industries.
- Lose economic value due to heavy licensing fees.

To change this trajectory, Indian companies must rethink their approach to R&D. They must move beyond short-term profits and invest in long-term technological innovation, ensuring that "Make in India" is supported by "Research in India."

4.3 Weak Industry-Academia Collaboration Compared to the West

In developed nations, industry and academia work together to drive technological advancements, product innovation, and economic growth. Universities provide cutting-edge research, skilled graduates, and experimental breakthroughs, while industries fund applied research, commercialization efforts, and large-scale implementation.

However, in India, this crucial partnership is severely underdeveloped. Instead of working closely with universities, most Indian corporations prefer to import technology or rely on foreign R&D investments. The result? A stagnant innovation ecosystem, where academic research remains theoretical, and industrial growth is dependent on external knowledge and imported technology.

This section explores why structured industry-academia collaborations are missing in India, how leading economies like the US, Germany, and China integrate research with industry needs, and the challenges preventing India from building a strong R&D-driven corporate culture.

4.3.1 The Lack of Structured Partnerships Between Universities and Corporations

Unlike developed countries, where companies actively collaborate with universities to drive innovation, India's industry-academia relationship is fragmented, weak, and often non-existent.

1. The Isolation of Universities from Industrial Needs

- Indian universities focus primarily on academic research, with very few projects aimed at real-world industrial applications.
- Most PhD research remains theoretical, with limited opportunities for commercialization.
- Unlike the US, where companies fund university R&D labs, Indian corporations rarely invest in university-based research projects.

2. Minimal Corporate Funding for University Research

- In developed nations, private companies fund over 50% of university research, while in India, over 90% of R&D funding comes from the government.
- Indian corporations prefer short-term technology imports instead of long-term investments in university research.
- In contrast, the US, Germany, and China provide tax incentives for companies to invest in university research, ensuring continuous collaboration and innovation.

3. Poor Internship and Talent Pipeline Integration

- Universities in India produce thousands of engineering, science, and management graduates every year, yet very few get hands-on industrial research experience.
- Unlike in the US and Germany, where students actively participate in corporate R&D projects, most Indian students graduate without exposure to real-world industrial research.
- The absence of industry-sponsored academic fellowships and incubators further weakens the research-to-market pipeline.

Without structured corporate funding and collaboration, university research in India remains disconnected from industrial needs, failing to produce commercially viable innovations.

4.3.2 How Global Models (US, Germany, China) Integrate Academic Research with Industrial Needs

The world's most advanced economies flourish because of strong industry-academia partnerships. They understand that research and innovation cannot happen in isolation—academic institutions and corporations must work together to create breakthroughs.

1. United States: The University-Industry Innovation Model

- The US leads in university-led industrial innovation, thanks to strong corporate investments in academic research.
- MIT, Stanford, and Harvard collaborate extensively with companies like Google, Tesla, and Microsoft, developing industry-relevant technologies.
- Government-backed programs like the Small Business Innovation Research (SBIR) initiative provide funding to connect university research with commercial enterprises.

2. Germany: The Research-Industry Dual Education System

- Germany's Fraunhofer Institutes serve as a model for industry-driven research, bridging universities and corporate needs.
- Companies like BMW, Siemens, and Bosch work directly with universities to develop next-generation automotive, automation, and engineering solutions.
- The government incentivizes R&D collaborations by offering tax benefits and shared research facilities for universities and corporations.

3. China: State-Led University-Industry Collaboration

- China's "Made in China 2025" initiative promotes university-industry partnerships in AI, semiconductors, and biotech.
- Companies like Huawei, Tencent, and Alibaba actively fund university research, driving rapid advancements in technology and AI.
- The Chincse government mandates that a portion of corporate profits be reinvested into university-led R&D, ensuring continuous industry-academia engagement.

These global models prove that when industries invest in university research, economies thrive. India must adopt similar collaborative strategies to break its dependence on foreign technology.

4.3.3 Challenges in Building a Research-Driven Corporate Culture in India

Despite the evident benefits of industry-academia collaboration, several structural and cultural barriers prevent Indian corporations from actively funding research.

1. Lack of Long-Term Vision Among Indian Businesses

- Many Indian companies focus on short-term profitability rather than long-term R&D investments.

- Unlike Samsung, Toyota, or Siemens, which invest billions in R&D, Indian corporations prioritize immediate production efficiency over future innovation.

2. Weak Government Incentives for Industry-Academia Collaboration

- Indian tax policies do not strongly encourage corporate R&D funding for universities.
- Government research grants are often bureaucratic and slow, discouraging private-sector participation.
- Unlike China and the US, India has no structured industrial R&D policy that mandates corporate-university partnerships.

3. Resistance to Risk-Taking and Innovation

- Indian industries tend to avoid high-risk R&D investments, preferring proven foreign technologies over indigenous development.
- Fear of failure and uncertain returns prevents companies from exploring high-potential emerging technologies.

4. Limited Knowledge Transfer from Academia to Industry

- In India, professors and researchers have little exposure to industrial problem-solving, limiting their ability to produce commercially viable solutions.
- Unlike in Germany and the US, where faculty members routinely consult for industries and launch startups, Indian professors are rarely involved in corporate innovation projects.

Without addressing these cultural and structural barriers, India will struggle to build a sustainable research-driven economy, continuing its reliance on foreign technology imports.

Conclusion

India's failure to establish strong industry-academia collaborations has crippled its ability to develop indigenous technologies. While global economies thrive on university-led industrial R&D, Indian universities remain isolated from corporate needs, and businesses neglect long-term research investments.

To bridge this gap, India must:

- Offer strong tax incentives for companies investing in university research.
- Create research incubators that connect corporations with academic institutions.
- Mandate industry-sponsored fellowships for university students to engage in industrial R&D.
- Encourage faculty participation in corporate innovation projects.

Unless these changes are implemented, Indian industries will remain technologically dependent on foreign companies, missing out on breakthrough innovations and economic self-sufficiency.

4.4 Case Studies: How India Missed Opportunities Due to Poor R&D Investment

India has long been celebrated as a manufacturing hub, a leader in generic pharmaceuticals, and a global IT services powerhouse. However, despite its strengths, the country has repeatedly failed to capitalize on key technological revolutions due to weak research and development (R&D) investments. While nations like China, the US, and Germany have poured billions into R&D, creating global tech giants, India has largely remained a consumer of foreign technology rather than a creator of new innovations.

This section examines four major industries where India had the potential to lead but missed out due to inadequate R&D investment—semiconductors, pharmaceuticals, electric vehicles, and AI/software development. These case studies illustrate how India's lack of long-term vision, risk-averse corporate culture, and weak government policies have prevented it from becoming a global innovation leader.

4.4.1 The Semiconductor Industry: Why India Failed to Develop Its Own Chip Manufacturing Ecosystem

Semiconductors are the backbone of modern technology, powering everything from smartphones and computers to AI systems and military defense equipment. Countries like the US, Taiwan, South Korea, and China dominate the global semiconductor market, but India remains entirely dependent on imports, despite having the capability to build its own chip manufacturing ecosystem.

Key Reasons for Failure

- **Lack of R&D Investment and Infrastructure:** India never invested in developing indigenous chip fabrication (fab) plants, despite having strong technical expertise. Setting up a semiconductor fab requires an initial investment of $5-10 billion, which neither the government nor private industry was willing to fund. Taiwan (TSMC) and South Korea (Samsung) heavily invested in fabs early on, securing their dominance in the global semiconductor market.
- **Delayed and Ineffective Government Policies:** India failed to implement a long-term semiconductor strategy, while China invested over $150 billion in its Made in China 2025 semiconductor plan. The 2021 Indian semiconductor

incentive policy came too late, and projects like the Tata Electronics fab in Dholera are still in early planning stages.
- **Dependence on Foreign Imports:**India imports more than 90% of its semiconductor needs from Taiwan, China, and the US.During the COVID-19 pandemic, semiconductor shortages crippled India's automobile and electronics industries, exposing the risks of import dependence.

Missed Opportunities and Consequences

- India could have developed a semiconductor hub decades ago, reducing its reliance on imports and positioning itself as a global supplier.
- The country continues to lag behind China, Taiwan, and South Korea, despite having the talent and demand to support a local industry.
- The lack of semiconductor manufacturing weakens India's national security, as modern defense systems depend on imported chips from foreign powers.

4.4.2 The Pharmaceutical Sector: How India Lags in Drug Discovery Despite Being a Generic Drug Giant

India is often referred to as the "Pharmacy of the World" due to its massive generic drug manufacturing industry. However, when it comes to original drug discovery and new molecule development, India is decades behind countries like the US, Germany, and China.

Key Reasons for Failure

- **Focus on Generics Rather Than Innovation:** Indian pharma companies focus almost entirely on producing generic drugs rather than developing new molecules. Unlike the US and Europe, where pharma giants like Pfizer, Novartis, and Roche invest heavily in R&D, Indian companies prioritize quick profits from generic drugs.

- **Lack of Government and Private Sector R&D Investment:** The US spends over 15% of its GDP on pharmaceutical R&D, while India spends less than 1%. Government R&D funding for drug discovery is minimal, and there is no structured policy to encourage innovation in the pharma sector.
- **Brain Drain and Talent Flight:** Top Indian researchers in drug discovery move to the US and Europe, where pharma R&D funding is significantly higher. India's best scientists work for foreign firms, creating new drugs that benefit global markets but not India.

Missed Opportunities and Consequences

- India could have been a leader in new drug discovery, developing innovative treatments for cancer, diabetes, and infectious diseases.
- The lack of original drug patents forces Indian firms to rely on licensing agreements with Western pharmaceutical giants.
- China has surpassed India in biopharma R&D, securing global patents and attracting billions in funding for biotech startups.

4.4.3 Automobile and Electric Vehicles: How India Lost Ground to China Due to Weak R&D Investment

The global automobile industry is undergoing a transformation with the rise of electric vehicles (EVs). While India had the potential to lead this shift, China has taken over the global EV market, leaving India far behind.

Key Reasons for Failure

- Slow R&D Investment in EV Technology: Indian companies like Tata Motors and Mahindra took too long to invest in EV R&D, while China poured billions into battery technology and EV production. BYD and CATL, Chinese firms, dominate the EV battery market, while Indian companies still rely on importing battery cells.
- Lack of Government Support for EV R&D: China provided subsidies, tax breaks, and incentives for local EV manufacturers, while India's FAME policies came much later and lacked funding. The absence of local EV research facilities forces Indian companies to buy battery and motor technology from foreign firms.
- No Long-Term Strategy for Domestic EV Manufacturing: Unlike China, which established EV research hubs, India lacks dedicated EV innovation centers. Most Indian EV makers assemble imported components instead of developing homegrown battery, motor, and AI technologies.

Missed Opportunities and Consequences

- India could have led the global EV market, reducing fuel dependence and becoming a major exporter.
- Instead, China dominates EV production, battery technology, and global sales, while India struggles with supply chain constraints.
- The lack of R&D investment has forced Indian automakers to depend on Chinese suppliers, weakening India's energy independence.

4.4.4 AI and Software Development: How India Remains a Service Hub Rather Than a Global Tech Innovator

India has been a leader in IT services and software outsourcing, but when it comes to cutting-edge AI, machine learning, and deep tech, the country is far behind the US and China.

Key Reasons for Failure

- **Lack of AI and Deep Tech R&D:** Indian IT firms focus on outsourcing and service-based solutions rather than AI-driven innovation. China, the US, and Europe invest heavily in AI research, quantum computing, and cybersecurity, while India lags in funding and policy support.
- **Minimal Government Investment in AI Startups:** The Indian government does not fund AI research at the scale seen in China and the US. While Google, Microsoft, and OpenAI lead AI breakthroughs, Indian AI startups struggle for funding and infrastructure.
- **Brain Drain of AI Talent:** India produces top AI engineers, but they move abroad for better research opportunities. Major AI breakthroughs come from Indian-origin scientists working in the US and UK, not India itself.

Missed Opportunities and Consequences

- India could have been a global AI leader, but its weak R&D culture limits cutting-edge innovations.
- Most Indian companies use foreign AI tools rather than building proprietary AI models.

Conclusion

India has missed multiple opportunities to become a global leader due to poor R&D investment and weak industry-government collaboration. Without a strong research culture, India will remain dependent on foreign technology, preventing true economic and technological independence.

5. The Talent Drain: Why Indian Researchers Go Abroad

India's failure to build a research-driven industrial and academic ecosystem has led to a mass exodus of talented scientists, engineers, and researchers to countries that offer better funding, superior infrastructure, and greater career opportunities. The previous chapters have highlighted India's lack of R&D investment, weak industry-academia collaboration, and excessive dependence on foreign technology—all of which contribute to a situation where India's brightest minds find more promising careers abroad than at home.

For decades, Indian researchers have migrated to the US, Europe, and even China, where well-funded laboratories, structured research grants, and strong industry partnerships provide them the resources and freedom to innovate. Instead of leading global scientific advancements from within India, these researchers contribute to Nobel-winning projects, AI breakthroughs, medical innovations, and space exploration—but for foreign institutions.

At the same time, India has failed to attract foreign researchers, further isolating its research ecosystem. Rigid bureaucracy, poor funding, and limited international collaboration make India an unattractive destination for global scientific talent. As a result, the country faces a double disadvantage—losing its best minds to foreign countries while failing to bring in international expertise.

This chapter explores why Indian researchers go abroad, how they succeed in foreign research environments, why India struggles to retain and attract global talent, and the long-term consequences of brain drain on India's scientific progress. Without urgent reforms, India risks remaining a country that produces talent only to serve the research ambitions of other nations.

5.1 Better Funding, Infrastructure, and Opportunities Overseas

Indian researchers, despite their talent and potential, increasingly migrate to foreign institutions in search of better funding, advanced research infrastructure, and career growth opportunities. Countries like the US, Germany, Canada, and China provide structured financial support, cutting-edge laboratories, and globally connected research networks, making them highly attractive for Indian scientists.

In contrast, India's weak R&D investment, bureaucratic hurdles, and lack of career progression discourage homegrown researchers from staying. The result is a massive talent drain, where India invests in education but fails to retain its top intellectual capital, leading to an over-reliance on foreign technological advancements.

This section explores the key reasons why Indian researchers prefer foreign institutions, highlighting how better funding, infrastructure, research freedom, and global collaboration opportunities contribute to this ongoing brain drain.

5.1.1 The Research Funding Gap: How India's Low R&D Investment Pushes Scientists Abroad

One of the biggest reasons Indian researchers leave the country is the severe lack of funding for scientific research. While developed nations invest a significant percentage of their GDP into R&D, India lags behind, offering limited grants, fewer fellowships, and inadequate financial support for long-term projects.

1. India's R&D Investment vs. Other Nations

- India spends less than 0.7% of its GDP on R&D, compared to: United States: 2.8%, China: 2.4%, Germany: 3.2%, South Korea: 4.5%
- The lack of funding directly affects scientists, research projects, and the ability to build world-class labs.

2. Limited Research Grants and Funding Restrictions

- Research funding in India is largely controlled by government bodies like CSIR, DST, DBT, and ICMR, which have limited budgets and slow approval processes.
- Unlike in the US and Germany, where independent researchers can apply for private and public grants, Indian scientists must be affiliated with an institution to qualify for funding.
- The grant approval process is bureaucratic, leading to delays, underfunding, and rejection of innovative projects due to rigid criteria.

3. Corporate Disinterest in Funding Research

- Indian industries rarely invest in fundamental research, focusing on short-term business goals rather than long-term scientific advancements.
- In contrast, in the US, corporations fund over 50% of total R&D spending, supporting research in AI, biotech, renewable energy, and materials science.
- Without corporate-backed research grants, many Indian scientists seek opportunities abroad where both government and private sector funding are available.

5.1.2 World-Class Research Labs and Facilities: Why Foreign Institutions Attract Indian Talent

India lacks the advanced research infrastructure necessary to support cutting-edge discoveries, forcing researchers to seek better-equipped institutions abroad.

1. The Gap in Research Infrastructure

- Leading universities in the US, Germany, and China invest billions in state-of-the-art laboratories, AI computing clusters, biotech research centers, and space research hubs.
- India's research facilities are outdated, with limited access to high-end equipment, making it difficult for scientists to conduct world-class experiments.

2. The Struggle for Access to Essential Research Tools

- Indian researchers lack access to high-performance computing facilities, genomic sequencing labs, and nanotechnology centers—all of which are easily available in foreign institutions.
- Many Indian universities still function with outdated lab equipment, and researchers often wait months for approvals to use specialized facilities.
- In the US and Germany, scientists have on-demand access to advanced research tools, allowing for faster experimentation and higher-quality results.

3. India's Dependence on Foreign Research Institutions

- Due to inadequate facilities, Indian researchers often collaborate with foreign universities for experiments requiring specialized instruments and funding.
- Instead of building its own research hubs, India continues to send top talent abroad, reinforcing its dependency on foreign advancements.

5.1.3 Higher Salaries and Career Growth: Financial and Professional Incentives for Researchers Abroad

Indian researchers earn significantly lower salaries compared to their counterparts in the US, Canada, Europe, and China, making foreign opportunities much more appealing.

1. The Salary Disparity Between India and Foreign Research Institutions

- An entry-level research scientist in India earns between ₹6-10 lakh per annum ($7,000-$12,000), while in: The US: $50,000–$100,000 per annum, Germany: €45,000–€80,000 per annum, China: ¥300,000–¥800,000 per annum
- This massive salary gap discourages top talent from staying in India, as researchers can earn 5–10 times more abroad.

2. Limited Career Progression and Job Stability

- Indian universities and research centers have poor career progression structures, making it difficult for young scientists to secure permanent positions or funding for independent research.
- In contrast, foreign institutions offer structured career paths, tenure-track positions, and better research funding support, ensuring long-term job stability.

5.1.4 Freedom to Innovate: How Less Bureaucracy and Better Research Autonomy Encourage Indian Talent to Leave

1. Bureaucratic Barriers to Research in India

- Research in India is highly bureaucratic, requiring multiple approvals for funding, lab access, and publication rights.

- The multi-layered hierarchy of Indian research institutions discourages young scientists from pursuing breakthrough ideas.
- In contrast, the US, UK, and Germany provide autonomy to researchers, allowing them to choose their own research directions without excessive administrative interference.

2. The Freedom to Pursue High-Risk, High-Reward Research

- Many breakthrough scientific discoveries come from high-risk, experimental research, which India rarely funds.
- Foreign institutions support cutting-edge fields like AI, quantum computing, and synthetic biology, while Indian funding agencies prefer safer, low-risk projects.

5.1.5 Global Research Collaborations: Why Indian Researchers Prefer Foreign Institutions with Strong International Networks

Top scientists thrive in environments that offer international collaboration opportunities, giving them access to the best minds, the latest research, and global funding networks.

1. The Strength of Global Research Networks

- Countries like the US, Germany, and China encourage multi-national collaborations, funding cross-border projects in AI, physics, medicine, and space exploration.
- India lacks structured international collaboration programs, limiting its researchers' exposure to global scientific advancements.

2. The Advantage of Publishing in High-Impact Journals

- Many high-impact journals prefer research from globally

recognized institutions, making it easier for Indian scientists abroad to publish.
- Foreign collaborations increase research credibility, allowing Indian-origin researchers to gain faster recognition and career growth.

Conclusion

India's failure to provide competitive research funding, infrastructure, salaries, and career opportunities is the primary reason why thousands of Indian researchers migrate to foreign institutions every year. Better-equipped laboratories, higher salaries, research autonomy, and strong international networks ensure that Indian scientists thrive abroad but struggle in India.

To reverse this brain drain, India must:

- Significantly increase R&D funding and infrastructure investment.
- Offer globally competitive salaries and structured career growth.
- Reduce bureaucracy and allow greater research autonomy.
- Encourage strong international collaborations and industry-academia partnerships.

Without these reforms, India will continue to lose its best minds, ensuring that scientific breakthroughs and technological leadership remain concentrated in foreign institutions.

5.2 Case Studies of Indian Researchers Succeeding Abroad

India has produced some of the world's most brilliant scientists, engineers, and researchers, yet many of them have achieved their greatest breakthroughs outside India. The lack of research funding, inadequate infrastructure, and slow career growth in India has pushed its top intellectual talent to foreign universities, research institutions, and corporate labs, where they have been able to excel and lead cutting-edge discoveries.

From pioneering work in artificial intelligence and medical sciences to leadership in space research and engineering, Indian-origin scientists have played a crucial role in shaping global innovation. This section highlights key case studies of Indian researchers who have thrived abroad, illustrating how India's failure to retain its top talent has benefited foreign institutions and industries.

5.2.1 The AI and Tech Brain Drain: How Indian-Origin Scientists Lead Global AI Research

Artificial Intelligence (AI) and advanced computing have become the foundation of modern technological progress, yet India has failed to emerge as a leader in AI innovation. Meanwhile, Indian-origin researchers and engineers are at the forefront of AI development in the US, UK, and Canada.

Key Examples of Indian-Origin Leaders in AI

- **Sundar Pichai (CEO, Google & Alphabet):** Under Pichai's leadership, Google has become a global AI powerhouse, leading breakthroughs in machine learning, natural language processing, and AI-driven search algorithms. While Pichai studied in India, his career growth happened in the US, where he had access to world-class research facilities and funding.
- **Pranav Mistry (AI and Augmented Reality, Former MIT Media Lab Researcher):** Known for his work on Sixth Sense Technology and AI-powered human-computer interaction, Mistry is a global innovator in AI-driven computing. He had to move to MIT to pursue his AI research, as India lacked the infrastructure for high-end AI experimentation.
- **Dr. Shivaram Kalyanakrishnan (AI & Robotics Research,**

Carnegie Mellon University & IIT Bombay): One of the leading researchers in reinforcement learning and robotics, he has significantly contributed to AI's application in autonomous systems. Despite returning to India to teach at IIT Bombay, he faces limited research funding compared to US institutions.

Why Indian AI Researchers Move Abroad

- Better funding for AI research in the US and Europe.
- Stronger industry collaborations with companies like Google, Tesla, and IBM.
- Access to advanced supercomputing and AI infrastructure.
- Higher salaries and career growth in AI-driven economies.

5.2.2 Nobel Laureates and Pioneering Scientists: Indian Minds Making Breakthroughs Overseas

Despite having some of the world's brightest scientists, India has seen many of its Nobel Prize-winning researchers thrive in foreign institutions.

Notable Indian-Origin Nobel Laureates in Science & Research

- Venkatraman Ramakrishnan (Nobel Prize in Chemistry, 2009): Ramakrishnan's groundbreaking work on ribosome structure was carried out in Cambridge, UK, not India. He has openly spoken about how Indian bureaucracy and lack of funding push scientists abroad.
- Abhijit Banerjee (Nobel Prize in Economics, 2019): Banerjee's work in poverty alleviation and behavioral economics was conducted at MIT, not an Indian research institution. He benefited from international research funding and advanced academic collaborations, both of which are rare

in India.

- Har Gobind Khorana (Nobel Prize in Medicine, 1968): A pioneer in genetic research, Khorana's Nobel-winning work was done in the US, despite being born and educated in India. India lacked biotech infrastructure, forcing him to move to the University of Wisconsin for advanced research.

Why India Fails to Retain Nobel-Worthy Talent

- Insufficient research funding and government support.
- Lack of high-end experimental labs in physics, medicine, and chemistry.
- No structured career paths for pioneering researchers.

5.2.3 Medical and Biotech Innovators: How Indian Researchers Are Advancing Healthcare Abroad

Indian-origin medical researchers have made game-changing contributions to global healthcare, but most of their breakthroughs happened outside India.

Key Examples of Indian-Origin Medical Researchers

- Dr. Siddhartha Mukherjee (Oncology & Cancer Research, Columbia University, US): Author of The Emperor of All Maladies, a Pulitzer Prize-winning book on cancer research. His work in oncology and genetics has influenced global cancer treatments, but India lacks the funding for such advanced research.
- Dr. Rakesh Jain (Biomedical Engineering, Harvard Medical School): A pioneer in tumor microenvironment research, Jain's work has led to new cancer treatment strategies. He works in the US because India lacks biotech infrastructure and clinical trial funding.

- Dr. Kiran Musunuru (Cardiovascular Genetics, Harvard & UPenn, US): A leading expert in gene-editing and heart disease research, his discoveries have helped develop new gene therapies. India does not support high-risk medical research, forcing many biotech scientists to move abroad.

Why Indian Biotech Researchers Leave

- Lack of world-class biomedical labs in India.
- Limited funding for genomics, stem cell, and biotech research.
- Difficult regulatory approvals for clinical trials in India.

5.2.4 Engineering and Space Research Leaders: How Indian Talent Contributes to NASA, CERN, and Leading Engineering Institutes

Many Indian-origin engineers and space scientists have played key roles in global scientific breakthroughs.

Key Examples

- **Dr. Kalpana Chawla (NASA Astronaut & Aerospace Engineer):** One of the most celebrated NASA astronauts, Chawla had to leave India due to limited aerospace research opportunities.
- **Dr. Subrahmanyan Chandrasekhar (Nobel Prize in Physics, 1983):** A key contributor to astrophysics and black hole research, but had to work at the University of Chicago for access to research funding.
- **Dr. Ajay Bhatt (Inventor of USB Technology, Intel, US):** Bhatt's work in USB and microchip technology shaped modern computing, but he had to work in Intel's US labs for technological development.

Why India Loses Its Engineering and Space Talent

- Limited opportunities in high-tech aerospace and defense research.
- Better funding at NASA, CERN, and MIT compared to ISRO and IITs.
- Slow bureaucratic processes and lack of risk-taking in Indian research institutions.

5.2.5 From India to the Ivy League: How Top Indian Students Choose Global Universities Over Indian Institutions

Every year, thousands of India's best students move abroad for higher education, especially to the Ivy League, MIT, Stanford, and Oxford.

Why Indian Students Choose Foreign Universities

- Better research funding and scholarships abroad.
- Access to world-class faculty and industry collaborations.
- Stronger career growth and global networking opportunities.

Impact of Losing Top Indian Students

- India invests in educating bright minds, but they contribute to foreign economies instead.
- The lack of research-driven universities in India forces students to look elsewhere for PhDs and postdoctoral studies.

Conclusion

These case studies prove that India produces world-class talent, but due to poor research funding, bureaucratic restrictions, and lack of infrastructure, its best scientists thrive in foreign institutions instead of Indian labs.

If India wants to retain and benefit from its intellectual capital, it must:

- Increase R&D funding to global levels.
- Develop world-class research institutions and labs.
- Create policies that encourage long-term innovation.

Without these reforms, India will remain a supplier of raw talent for foreign innovation, missing out on its own scientific and technological potential.

5.3 Why India Doesn't Attract Foreign Researchers

While India has a long history of producing top scientists who succeed in global research institutions, it struggles to attract foreign researchers to work in Indian universities, research labs, and corporate R&D centers. Unlike the US, Europe, and China, which actively recruit international talent, India remains isolated from global research networks, discouraging foreign scientists from contributing to its research ecosystem.

Even Indian-origin researchers who achieve global recognition abroad rarely return, due to limited funding, rigid bureaucracy, and a lack of professional incentives. India's failure to create an attractive research environment for foreign talent has led to a one-way brain drain, where India loses its best minds while failing to bring in global experts.

This section explores why India struggles to attract foreign researchers, analyzing the key barriers that prevent India from becoming a global research hub.

5.3.1 Limited Global Collaboration: Why Foreign Researchers Prefer the US, Europe, and China

Global research thrives on international collaborations, where universities, companies, and governments work together to develop breakthrough technologies. However, India has failed to integrate itself into the global research network, making it an unattractive destination for foreign scientists.

1. India's Weak Presence in Global Research Networks

- Leading global institutions in the US, UK, Germany, and China have structured programs to invite foreign scientists and integrate them into national research projects.
- India lacks an equivalent framework, making it difficult for foreign researchers to secure funding, lab access, or institutional backing.
- The US National Science Foundation (NSF), European Research Council (ERC), and Chinese Academy of Sciences (CAS) regularly fund international researchers, while India's Department of Science & Technology (DST) does not have a competitive global recruitment program.

2. The Absence of International Research Exchange Programs

- The US and European universities offer visiting fellowships to attract global scientists, while Indian institutions have few opportunities for international faculty and researchers.
- Countries like China, Germany, and Singapore actively invite top researchers by offering attractive grants and long-term positions.
- India's lack of institutional partnerships with global research bodies discourages foreign talent from considering Indian universities and research labs.

5.3.2 Rigid Bureaucracy and Visa Barriers: Challenges for Foreign Talent in Indian Research Institutions

Even when foreign researchers show interest in working in India, bureaucratic inefficiencies and visa restrictions make it extremely difficult for them to join Indian institutions.

1. Complex and Slow Research Visa Approval Process

- Unlike the US and Europe, where research visas are fast-tracked, India has a complicated and slow visa approval process for foreign researchers.
- Many scientists who apply for Indian research visas face months of delays, discouraging them from working in India.
- China has introduced a "Fast-Track Visa for Foreign Scientists", making it easier for global researchers to enter the country, while India still lacks a streamlined system.

2. Rigid Employment Rules for Foreign Academics

- Indian universities have strict hiring requirements, making it difficult for foreign researchers to obtain long-term positions.
- Unlike Western institutions, where professors and scientists are hired based on merit, Indian universities require extensive government approvals for foreign hires.
- Many foreign researchers lose interest due to unclear and unpredictable hiring procedures in Indian institutions.

5.3.3 Lack of Competitive Salaries and Grants: Why India Fails to Attract and Retain Top Global Talent

Attracting global researchers requires competitive salaries, substantial grants, and well-funded laboratories. However, India fails to offer financial incentives that would encourage international scientists to choose Indian institutions over the US, Europe, or China.

1. Salary Comparison: India vs. Global Research Institutions

In the US, Europe, and China, researchers receive highly competitive salaries:

- US (MIT, Stanford, Harvard): $90,000–$150,000 per year
- Germany (Max Planck Institutes, TU Munich): €80,000–€120,000 per year
- China (Tsinghua University, CAS): ¥600,000–¥1,200,000 per year
- India (IISc, IITs, CSIR): ₹12–₹20 lakh per year ($15,000–$25,000)

Indian salaries for top researchers are less than 20% of what leading global institutions offer, making India an unattractive option for foreign talent.

2. Limited Research Grants and Funding

- Indian institutions offer fewer and smaller research grants compared to global competitors.
- The US NSF, China's National Natural Science Foundation, and the European Research Council offer grants 10–15 times higher than what India provides.

Without large-scale grants, research autonomy, and industrial sponsorships, India cannot compete for top international talent.

5.3.4 No Industry-Driven Research Ecosystem: How Weak Corporate R&D Discourages Foreign Researchers

In leading economies like the US, Germany, and China, corporations actively fund university research and recruit global scientists for industry-driven projects. However, in India, corporate

R&D remains weak, providing few opportunities for foreign researchers to collaborate with industry.

1. Lack of Corporate-Backed Research Positions

- In the US, major tech firms like Google, Microsoft, and Tesla fund research fellowships at MIT, Stanford, and UC Berkeley, attracting the world's top researchers.
- In contrast, Indian corporations rarely invest in fundamental research, leading to fewer industry-backed opportunities for foreign talent.

2. No Start-Up or Innovation-Led Research Culture

- The US, Israel, and Germany attract researchers through highly funded start-ups and venture capital-driven R&D.
- India's start-up ecosystem remains focused on IT services and e-commerce, with limited investment in deep-tech research and innovation.

Without strong industry-research collaboration, India remains unappealing to foreign scientists who prefer innovation-driven ecosystems like Silicon Valley, Beijing, and Berlin.

5.3.5 Cultural and Institutional Barriers: Why International Researchers Find It Hard to Integrate into India's Research System

Even if foreign researchers overcome visa, salary, and funding challenges, they often face cultural and institutional difficulties that make working in India frustrating.

1. Rigid Academic Hierarchy and Bureaucratic Control

- Indian research institutions have a strict hierarchy, where junior researchers and foreign scientists often struggle to get equal recognition.

- Unlike in the US and Europe, where universities encourage flat, open academic discussions, India follows a seniority-based system, limiting opportunities for young international researchers.

2. Limited Freedom for Independent Research

- Foreign researchers are used to flexible, innovation-driven research environments in the West.
- In India, bureaucracy and rigid research guidelines prevent them from pursuing high-risk, high-reward scientific experiments.

3. Language and Cultural Integration Issues

- Many Indian research institutions do not have structured programs to help foreign scientists adapt.
- In contrast, China has established cultural integration programs and language support, making it easier for foreign researchers to work and collaborate.

Conclusion

India's failure to attract foreign researchers is a result of limited funding, excessive bureaucracy, weak industry collaboration, and lack of international research networks. In contrast, countries like the US, Germany, and China actively recruit foreign scientists by offering better salaries, easier visa policies, and a strong research culture.

- If India wants to become a global research hub, it must:
- Increase research funding and salaries to international standards.
- Streamline visa approvals and hiring processes for foreign

researchers.

- Encourage industry-academia collaboration to create research-driven jobs.
- Build structured international exchange programs and partnerships.

Without these changes, India will continue to lose its best minds while failing to attract global talent, keeping it on the sidelines of global scientific advancements.

5.4 The Impact of Brain Drain on India's Scientific and Technological Growth

India's brain drain problem is not just about individual researchers leaving for better opportunities—it has far-reaching consequences for India's scientific progress, industrial self-sufficiency, and global technological leadership. When India's top engineers, scientists, and researchers migrate abroad, the country loses intellectual capital, weakens its innovation ecosystem, and becomes increasingly dependent on foreign technologies.

Despite having some of the world's best scientific minds, India often fails to lead in global breakthroughs, as its top researchers contribute to foreign economies rather than driving domestic advancements. The lack of high-end research, slow industrial growth, and missed leadership in emerging technologies are all symptoms of this ongoing crisis.

This section explores the four major impacts of brain drain on India's scientific and technological growth, highlighting the urgent need for reforms to retain and attract top talent.

5.4.1 Loss of Intellectual Capital: How Brain Drain Affects India's Innovation Potential

India invests heavily in educating scientists and engineers, yet a significant portion of its top talent leaves the country permanently.

This results in a loss of intellectual capital, where India's brightest minds drive research and innovation abroad while domestic institutions and industries struggle with a talent shortage.

1. India's Best Minds Are Powering Foreign Institutions

- Indian-origin scientists and engineers hold leadership positions in NASA, Google AI, MIT, Stanford, and global biotech firms.
- Top Indian researchers are making breakthroughs in cancer research, space exploration, AI, and quantum computing—but in the US, UK, and Germany, not in India.

2. The "Skill Drain" in Cutting-Edge Fields

- The most highly skilled AI researchers, space scientists, and biotech experts prefer to work in the US, Canada, and Europe, where they have access to better infrastructure and research freedom.
- India struggles to build world-class R&D teams, forcing it to import foreign experts at a higher cost.

3. The Innovation Gap Between India and the West

- While China and the US consistently produce Nobel-winning research, India lags in global patents and breakthrough discoveries.
- Indian institutions focus on incremental research rather than disruptive innovation, partly due to the loss of top scientific minds to foreign universities and companies.

Unless India finds ways to retain its best researchers, it will continue to lose intellectual capital and struggle to create groundbreaking scientific advancements.

5.4.2 Dependence on Foreign Technology: Why India Relies on Global Discoveries Instead of Leading Them

One of the biggest consequences of brain drain is India's increasing dependence on foreign technology. Instead of leading global innovation, India remains a consumer of foreign patents, scientific discoveries, and industrial breakthroughs.

1. Over-Reliance on Imported Technology

- India imports more than 80% of its semiconductor needs, mainly from Taiwan and China.
- The Indian defense sector relies on Russian, Israeli, and American technology, instead of building indigenous systems.
- Even in biotechnology and pharmaceuticals, India remains a manufacturer of generic drugs, while real innovation happens in Western labs.

2. Delayed Adoption of New Scientific Breakthroughs

- Since India does not produce high-tech innovations, it is forced to wait for global advancements to be commercialized before adoption.
- Example: India lagged in developing AI-based healthcare solutions while US and European firms advanced precision medicine and genomics research.

3. Economic Cost of Importing Technology

- India spends billions of dollars annually in licensing fees, royalties, and technology transfers, which could have been saved if it had homegrown innovation capabilities.
- Instead of being a global exporter of intellectual property,

India continues to buy technology from other nations, creating a long-term economic disadvantage.

If India does not develop its own technological ecosystem, it will remain dependent on foreign research, limiting its ability to compete in high-tech industries.

5.4.3 Missed Leadership in Emerging Technologies: How AI, Biotech, and Quantum Computing Suffer Due to Talent Migration

The next global technological revolutions will be driven by artificial intelligence, biotechnology, and quantum computing. However, due to brain drain, India is not leading in any of these fields.

1. The AI Leadership Gap

- US companies like OpenAI, Google, and Facebook lead global AI development, employing Indian-origin AI experts who should have been working in India.
- China has invested billions in AI research, while India lacks dedicated AI research hubs.
- India's best AI minds are working in Silicon Valley and Canadian AI labs, rather than developing India's own machine learning ecosystem.

2. The Biotech and Pharma Setback

- Despite being the world's largest generic drug producer, India lags behind in biotechnology R&D.
- Indian biotech researchers have migrated to the US, UK, and Germany, where they develop cancer treatments, stem cell therapy, and genetic medicine.
- Countries like China have already built national biotech hubs, while India still lacks proper funding for biotech innovation.

3. The Quantum Computing Disadvantage

- The US, China, and Europe are leading in quantum computing, investing billions in developing quantum processors and security systems.
- India has very few researchers working in quantum mechanics, as top physicists prefer to work at MIT, Harvard, or European quantum research institutes.
- Without expertise in quantum computing and cybersecurity, India risks falling behind in next-generation computing and cryptography.

The future of global power will be determined by who dominates AI, biotech, and quantum computing—if India does not reverse its brain drain, it will remain a secondary player in these critical industries.

5.4.4 Economic Consequences: The Financial Cost of Educating Researchers Who Later Contribute to Foreign Economies

Brain drain is not just a scientific loss—it is also an economic burden. The Indian government and taxpayers invest billions in educating top scientists and engineers, but other countries benefit from their contributions.

1. The Cost of Educating Scientists Who Leave India

- IITs, AIIMS, and IISc produce some of the world's best scientists, yet many graduates move abroad, meaning India gets no return on its investment.
- The Indian government spends ₹10–15 lakh per student at IITs, but nearly 40% of IIT graduates move abroad for jobs or higher studies.

2. Brain Drain's Contribution to Foreign Economies

- Indian-origin scientists generate billions in research patents and industrial profits for the US, UK, and Germany.
- For Example: Indian engineers at NASA, Google, Tesla, and Microsoft drive multi-billion-dollar innovations that benefit foreign economies.

3. The Need for Retention Policies to Protect Economic Investment

- Countries like China have policies to bring back top scientists, offering high salaries, grants, and world-class research facilities.
- India lacks structured incentives to attract its best minds back, leading to a continuous outflow of talent.

If India does not find ways to retain its intellectual capital, it will continue losing its best scientists to foreign economies, ultimately weakening its global economic position.

Conclusion

Brain drain is one of the biggest threats to India's scientific and technological growth. The loss of intellectual capital, dependence on foreign technology, missed leadership in key fields, and economic setbacks have slowed India's rise as a global innovator.

To reverse this trend, India must:

- Increase research funding to retain top scientists.
- Build world-class laboratories and industry-driven R&D hubs.
- Offer competitive salaries, grants, and career growth opportunities.
- Develop policies to attract Indian-origin researchers back to India.

If India does not take immediate action, it will continue to lose its brightest minds, ensuring that scientific and technological leadership remains in the hands of other nations.

5.5 Reversing the Brain Drain: How India Can Retain and Attract Top Researchers

India's brain drain crisis has left the country struggling to retain its brightest minds in science, technology, and engineering. While other nations actively attract and integrate top researchers into their innovation ecosystems, India has largely failed to create an environment conducive to high-quality research and scientific discovery.

Reversing brain drain is not just about stopping Indian scientists from leaving—it is about making India a preferred destination for research and innovation. The solution lies in increasing research funding, building world-class institutions, reforming policies for returning scientists, strengthening industry-academia partnerships, and fostering international collaborations.

This section outlines the key reforms India must implement to retain its talent, attract Indian-origin researchers back, and position itself as a global research powerhouse.

5.5.1 Increasing Research Funding and Infrastructure: The Need for More Government and Corporate Investment

One of the biggest reasons Indian researchers leave is the severe lack of funding and poor infrastructure in universities and research institutions. To retain and attract top researchers, India must significantly increase R&D investment and build state-of-the-art laboratories and technology hubs.

1. Raising India's R&D Expenditure to Global Standards

- India currently spends less than 0.7% of GDP on R&D, compared to: US: 2.8% of GDP, Germany: 3.2% of GDP, China: 2.4% of GDP, South Korea: 4.5% of GDP
- Increasing R&D investment to at least 2% of GDP would make India more competitive in global innovation.

2. Creating Research Superclusters and Tech Parks

India needs to develop research superclusters similar to:

- Silicon Valley (US) for AI and tech startups.
- Zhongguancun (China) for hardware and quantum computing.
- BioValley (Germany) for biotechnology.

These hubs should provide:

- State-of-the-art laboratories and research facilities.
- Access to high-performance computing, biotech labs, and AI centers.
- Direct funding from both the government and private investors.

3. Increasing Private Sector Investment in R&D

Indian companies invest very little in fundamental research. In contrast:

- Google, Microsoft, and IBM fund AI and quantum computing research.
- Pfizer and Novartis fund global biotech innovations.
- Tesla and Samsung fund battery and EV research.

India must introduce tax incentives and grants to encourage corporations to fund university research and tech startups. Without substantial increases in funding and infrastructure, India will continue to lose its top researchers to better-equipped foreign institutions.

5.5.2 Creating World-Class Research Institutions: How India Can Compete with Global Universities

India has some prestigious institutions like IITs, IISc, and AIIMS, but they fail to compete with the world's top research universities in funding, faculty, and research output. To become a global research hub, India must upgrade its universities and create new world-class institutions.

1. Establishing India's Own Ivy League-Style Research Universities

India must create globally recognized, well-funded research universities, similar to: MIT, Stanford, Harvard (US), ETH Zurich (Switzerland), Tsinghua University (China).

These institutions should:

- Offer competitive salaries to retain the best faculty.
- Have strong funding for research projects.
- Partner with global universities and industries for technology transfer.

2. Introducing Flexible Academic Structures

Unlike rigid Indian universities, global institutions allow:

- Tenure-track systems that provide long-term career security.
- Startup incubation within universities, where students and professors can launch tech companies.
- Independent research funding for faculty without bureaucratic restrictions.

3. Attracting International Faculty and Researchers

Top research institutions attract global talent, while Indian universities have almost no foreign faculty.

Policies should be introduced to:

- Ease visa restrictions for international professors.
- Offer competitive salaries for global researchers.
- Provide funding for joint research projects with foreign universities.

India cannot retain its own talent if it does not create world-class research institutions. Without urgent reforms, Indian students and researchers will continue to choose Harvard, Oxford, and Tsinghua over IITs and IISc.

5.5.3 Policy Reforms for Returning Scientists: Offering Incentives for Indian Researchers to Come Back

Many Indian-origin scientists want to return to India but are discouraged by low salaries, lack of infrastructure, and bureaucratic hurdles. To reverse brain drain, India must introduce structured incentives for returning scientists.

1. "Brain Gain" Programs for Returning Scientists

Countries like China and Israel have successfully brought back thousands of top researchers through targeted policies:

China's "Thousand Talents Plan" provides:

- $500,000 grants for returning scientists.
- Guaranteed research funding.
- State-of-the-art lab facilities.

Israel's Yozma Program offers:

- Tax incentives for returning researchers.
- Research partnerships with global institutions.

India must introduce a similar "Returning Scientists Program" that includes:

- High-value research grants (₹5-10 crore per scientist).
- Fast-track research project approvals.
- Assured faculty positions in top Indian institutions.

2. Reducing Bureaucratic Barriers for Returning Scientists

- Many Indian scientists face bureaucratic delays when applying for research funding or academic positions.
- India must introduce a "Fast-Track Research Approval System", ensuring returning scientists get immediate access to funding and infrastructure.

5.5.4 Building a Strong Industry-Academia Link: How Research Commercialization Can Keep Talent in India

The best researchers want their discoveries to be commercialized and used in real-world applications. However, India's weak industry-academia collaboration discourages top scientists from staying.

1. Encouraging Research Commercialization

- US universities turn research into billion-dollar companies (Google, Facebook, Tesla).
- Indian universities fail to commercialize innovations, causing researchers to leave for better startup ecosystems abroad.
- **Solution:** Set up University-Industry Research Centers that help researchers turn ideas into commercial products.

2. Incentives for Industry-Funded Research

- India must introduce corporate tax breaks for companies that fund university R&D.
- Industries should be required to fund a portion of their R&D through university collaborations, ensuring long-term research funding.

5.5.5 Encouraging International Collaborations: Making India a Research Hub for Global Scientists

To attract top global researchers, India must position itself as a key player in international science and technology.

1. Creating Global Research Partnerships

- Establish joint research programs with MIT, Stanford, Cambridge, and Tsinghua.
- Set up India-based international research centers to invite global scientists.

2. Hosting World-Class Research Conferences

- India rarely hosts major global science and technology summits.
- Organizing global AI, biotech, and physics conferences would position India as a leader in cutting-edge research.

Conclusion

Reversing brain drain requires urgent policy reforms in funding, infrastructure, academic flexibility, industry partnerships, and international collaborations.

To retain and attract top scientists, India must:

- Increase research funding to match global standards.
- Establish world-class research universities and tech hubs.
- Offer high-value incentives to returning Indian scientists.

- Strengthen industry-academia partnerships for research commercialization.
- Position India as a global research hub through international collaborations.

If these changes are not made, India will continue to lose its brightest minds, ensuring that scientific leadership remains in the hands of other nations.

6. The Challenges of Publishing and Recognition

Despite the intellectual potential and dedication of many Indian researchers—both institutional and independent—getting their work published and recognized remains an uphill battle. The previous chapter explored how brain drain and inadequate research funding force India's brightest minds to seek opportunities abroad. However, even those who choose to stay in India and contribute to research face significant challenges in getting published, gaining recognition, and securing credibility in the global academic community.

Academic publishing is heavily biased toward institutional research, making it difficult for independent researchers and scholars from underfunded institutions to break through. Many top-tier journals and conferences prioritize research from well-funded universities, leaving independent scholars struggling for visibility. The peer review system often favors established academic networks, creating a gatekeeping effect that makes it difficult for new or unaffiliated researchers to get published.

Adding to the problem, predatory journals exploit independent researchers, charging exorbitant fees for publication without proper peer review, trapping many scholars in a cycle of low-impact, unrecognized work. Even when independent research gets published in legitimate journals, citation bias further limits its recognition, as elite universities and institutions dominate academic references and networking circles.

This chapter explores the systemic challenges in academic publishing, focusing on institutional bias, the struggles of independent researchers, peer review challenges, predatory journals, and citation bias. It also discusses practical solutions that can help independent researchers overcome these barriers, build global networks, and gain the recognition they deserve.

6.1 Institutional Bias in Academic Publishing

Academic publishing is the backbone of scientific recognition and progress, yet it remains heavily biased toward institutional research. While universities and research institutions provide structured support, funding, and peer networks, independent researchers struggle to gain entry into the system. This institutional favoritism creates a significant gap between affiliated and unaffiliated scholars, making it difficult for independent researchers to publish in top-tier journals and participate in high-impact conferences.

The result? Many high-quality, innovative research works from independent scholars are overlooked, while institution-backed research—regardless of its merit—often gets prioritized due to established academic networks, funding support, and journal prestige metrics.

This section examines how institutional bias affects publishing, why journals and conferences prefer university-backed research, and how institutional funding increases the likelihood of publication success.

6.1.1 Why Journals and Conferences Prefer Institutional Affiliations Over Independent Research

Publishing in a high-impact journal is often seen as the gold standard of research recognition. However, most journals and academic conferences favor submissions from researchers affiliated with recognized institutions, creating a major disadvantage for independent scholars.

1. Reputation and Credibility Bias in Publishing

- Many top journals and conference committees assume that institution-backed research is more credible because it comes from universities with established academic reputations.
- Independent researchers, even if their work is groundbreaking, are often seen as outsiders to the academic system, making it harder for them to gain trust and acceptance.
- Reviewers, often from elite institutions, may unconsciously favor research from colleagues or institutions they are familiar with.

2. Institutional Backing Provides Access to Established Academic Networks

- Institutional researchers benefit from collaborations, mentorships, and reference networks, which make it easier to get published in top-tier journals.
- Many high-impact journals have unwritten norms where papers authored by researchers from Harvard, MIT, Oxford, or IITs receive quicker attention than those from unknown or unaffiliated scholars.
- Independent researchers often struggle to find co-authors or references from established academics, which weakens their chances of acceptance.

3. Conference Selection Committees Favor Institutional Researchers

- Major academic conferences (e.g., IEEE, ACM, Nature Conferences, NeurIPS) give preference to submissions from renowned universities and research institutes.
- Institutional researchers are often invited as speakers and panelists, while independent researchers struggle to get their papers accepted or even reviewed.
- Even early-stage, low-quality papers from top universities often get accepted, while high-quality independent research faces stricter scrutiny.

This institutional bias makes it significantly harder for independent researchers to gain recognition in the academic world, no matter how valuable their research is.

6.1.2 How University-Backed Researchers Get Easier Access to Top-Tier Journals

Publishing in top-tier journals (e.g., Nature, Science, The Lancet, IEEE Transactions, JAMA) is a crucial milestone for any researcher. However, institution-backed researchers have a distinct advantage in this highly competitive space.

1. Institutional Branding Enhances Journal Acceptance Rates

- Many high-impact journals are associated with elite universities and research institutions, leading to a systemic preference for papers from these organizations.
- A researcher from MIT or Stanford will often find it easier to get their paper reviewed and published than an independent researcher with no formal affiliation.
- Even editorial boards and peer reviewers are often drawn from major institutions, reinforcing a cycle of institutional favoritism.

2. Access to Top-Tier Journal Submissions Through Faculty Networks

- Institutional researchers can submit their papers through professors and senior academics who already have established relationships with journals.
- Many top-tier journals invite researchers for publications, a privilege that independent researchers do not have.
- University researchers also receive guidance from faculty members on structuring papers in a way that maximizes acceptance chances.

3. Institutional Support in Navigating Journal Submission and Peer Review

Many universities have dedicated publication support teams that help researchers with:

- Formatting and structuring their papers according to journal guidelines.
- Addressing peer review feedback and making necessary modifications.
- Resubmitting rejected papers with strategic adjustments.
- Independent researchers lack access to such support, making the complex peer review and submission process even more difficult.

Even if an independent researcher produces high-impact work, they often struggle to get past the editorial review stage, while institution-backed researchers receive assistance, mentorship, and fast-track pathways to journal publication.

6.1.3 The Role of Institutional Funding in Increasing Publication Chances

Academic publishing is not just about research quality—it is also deeply tied to funding. Institutions that invest heavily in research naturally produce more publications, giving their researchers an advantage.

1. Universities and Institutions Fund Research That Is More Likely to Get Published

Institutional funding helps researchers:

- Conduct large-scale experiments that independent researchers cannot afford.
- Purchase high-end laboratory equipment that increases the depth and credibility of their research.
- Hire research assistants and statisticians to refine their work before submission.

Journals and reviewers often favor studies with extensive experimental backing, which independent researchers struggle to achieve due to funding limitations.

2. Institutional Researchers Can Afford High Journal Submission Fees

Many high-impact journals charge hefty publication fees (often ranging from $2,000 to $10,000 per article). Institutional researchers can:

- Get these costs covered by university grants or department funding.
- Use institutional discounts and fee waivers.

Independent researchers must bear these costs personally, making it financially difficult to publish in top journals.

3. Universities Fund Conference Participation, Giving Institutional Researchers More Visibility

Attending top conferences (e.g., NeurIPS, ICML, SIGGRAPH, IEEE, ACS conferences) costs thousands of dollars in registration, travel, and accommodation fees.

Institutional researchers receive:

- Travel grants and conference sponsorships.
- Priority invitations to present research at major events.

Independent researchers often cannot afford to attend, losing opportunities to network and gain recognition in their field.

This financial advantage ensures that institutional researchers remain dominant in publishing, conferences, and citations, while independent researchers struggle to even enter the system.

Conclusion

The institutional bias in academic publishing creates a deeply unfair system where university-affiliated researchers have easier access to top journals and conferences, while independent researchers struggle to gain visibility and recognition.

Key barriers include:

- Journals and conferences prioritizing institution-backed research.
- University researchers benefiting from academic networks and faculty mentorship.
- Institutional funding giving affiliated researchers an advantage in research quality, submission fees, and conference participation.

To level the playing field, the academic community must:

- Create more open-access platforms that evaluate research on merit, not institutional affiliation.
- Reduce financial barriers to publication and conference

participation.
- Encourage independent research fellowships that support unaffiliated scholars.

Without these changes, independent researchers will remain locked out of top-tier publishing, and academic progress will continue to be dictated by institutional privilege rather than intellectual merit.

6.2 The Struggle of Independent Researchers to Gain Credibility

For any researcher, credibility is the foundation of academic success. A published paper in a reputable journal, citations in scholarly work, and invitations to conferences validate a researcher's contributions and influence their future opportunities. However, for independent researchers, establishing credibility is a major challenge, as the academic world tends to favor institutional affiliations over individual merit.

Without the backing of a university or research institute, independent scholars struggle to get their work accepted in top journals, secure funding for publication, and gain recognition for their contributions. Many breakthroughs by unaffiliated researchers go unnoticed, delayed, or outright rejected, not because of their lack of quality, but due to systemic biases in the academic publishing world.

This section explores how independent researchers struggle to gain credibility, examining the barriers they face in publishing, funding, and recognition.

6.2.1 How Independent Researchers Face Rejection Due to Lack of Affiliation

Most academic journals and conferences require researchers to list an institutional affiliation when submitting their work. This automatically puts independent scholars at a disadvantage, as many journals are hesitant to accept papers from unaffiliated researchers.

1. Institutional Affiliation as a Requirement for Publication

- Many high-impact journals prioritize submissions from established institutions, assuming that researchers from universities produce more credible and peer-reviewed work.
- Some journals reject submissions outright if an author does not list a university, research institute, or corporate R&D lab.
- Even if an independent researcher submits a high-quality paper, editors and reviewers often doubt the legitimacy of the work, assuming a lack of academic rigor.

2. Gatekeeping in the Peer Review Process

- The double-blind peer review system is supposed to remove bias, but in reality, journal editors often select reviewers from elite institutions, who may favor institution-backed research.
- Independent researchers frequently receive more scrutiny than their institutional peers, as reviewers may assume that a lack of affiliation means a lack of proper research methodology.
- Without an institutional network to advocate for their work, independent scholars face higher rejection rates, even when their research is groundbreaking.

3. The Preference for Established Academic Networks

- Many top-tier journals operate within exclusive academic circles, where research from Harvard, MIT, Oxford, or IITs gets immediate attention, while work from unaffiliated researchers is overlooked.
- University researchers benefit from faculty mentorship, which helps them structure their work to meet journal expectations, a resource independent scholars lack.

- Even when independent research is cited in other papers, it is often overlooked in mainstream discussions due to its lack of institutional backing.

Without a university name attached to their work, independent researchers struggle to convince journal editors, reviewers, and readers that their research is credible.

6.2.2 The Challenge of Funding Publication Fees Without Institutional Support

Even if an independent researcher successfully gets their paper accepted, they face another major hurdle—funding the publication. Many academic journals, particularly high-impact open-access journals, require researchers to pay hefty publication fees to get their work published.

1. High Publication Costs in Academic Journals

- Top journals charge publication fees ranging from $2,000 to $10,000 per article, creating a huge financial barrier for independent researchers.
- Institutional researchers can cover these costs through university funding, research grants, or departmental budgets, but independent researchers must pay out of pocket.
- Many independent scholars lack access to financial aid, fellowships, or research grants that institutional researchers receive.

2. Limited Access to Waivers and Discounts

- Some journals offer fee waivers or discounts, but these are usually reserved for researchers from developing countries or students.
- Independent researchers, particularly those from middle-

income backgrounds, often do not qualify for waivers, forcing them to either self-fund or publish in lower-tier journals.
- As a result, many independent scholars avoid submitting to top journals, limiting their research's impact and visibility.

3. The Burden of Self-Financed Research

Independent researchers must personally fund their entire research process, including:

- Experimentation costs (if applicable).
- Access to paid research databases, journals, and libraries.
- Travel expenses for conferences and networking events.

Institutional researchers receive financial support for all these aspects, while independent scholars face financial exhaustion just to get their work published.

Without institutional backing, the financial burden of publishing makes it nearly impossible for independent researchers to compete with university-backed scholars.

6.2.3 How Mainstream Journals Overlook Groundbreaking Work from Unaffiliated Scholars

Throughout history, some of the most important scientific discoveries have come from independent researchers. However, in modern academia, groundbreaking independent research is often ignored, delayed, or dismissed due to the systemic bias in publishing.

1. Historical Examples of Overlooked Independent Research

- Gregor Mendel's work on genetics was ignored by the scientific community for decades because he was not affiliated with a major university.
- Einstein's early work on special relativity was dismissed by many journals before gaining recognition because he was

working as a patent clerk, not a university professor.

- Ramanujan, the self-taught mathematician, faced skepticism until he was invited to Cambridge—an example of how institutional recognition determines the acceptance of ideas.

2. How Modern Independent Research Is Sidelined

- Many independent researchers work on interdisciplinary fields, unconventional theories, or emerging topics that mainstream academia often ignores until validated by institutions.
- Studies that challenge existing academic paradigms face more resistance in peer review, especially if they come from non-university researchers.
- Even when independent scholars publish in smaller journals, their work is rarely cited by institutional researchers, reducing its overall impact.

3. The Lack of Visibility and Academic Recognition

Citations and peer discussions help researchers build academic credibility, but independent researchers:

- Struggle to get their work cited in major studies.
- Are not invited to participate in academic panels, conferences, and symposiums.
- Have difficulty getting research funding due to a lack of previous high-impact publications.

Institutional researchers benefit from departmental networking, faculty collaborations, and conference invitations, giving them more visibility even when their work is less impactful.

This institutional dominance means that many breakthrough ideas from independent scholars remain hidden or take decades to gain recognition.

Conclusion

For independent researchers, the journey to academic recognition is full of obstacles. Even with high-quality work, they face:

- Rejection from top journals due to lack of institutional affiliation.
- The financial burden of self-funded research and expensive publication fees.
- A systemic bias in mainstream academia that overlooks groundbreaking independent research.

These challenges create a barrier between mainstream academia and independent scholars, making it difficult for outsiders to contribute to scientific progress.

To address this issue, academic publishing must:

- Adopt a more inclusive peer review process that evaluates research on merit, not affiliation.
- Provide financial aid and grants specifically for independent researchers.
- Encourage open-access platforms that recognize and promote non-institutional research.

Without such reforms, academia risks losing valuable contributions from independent scholars, delaying innovation and reinforcing institutional elitism.

6.3 The Disconnect Between Policy and Real Research Needs

Academic publishing and research funding should ideally promote innovation, diverse perspectives, and groundbreaking discoveries. However, in reality, the policies governing research funding and publishing often fail to support independent researchers, favoring institution-backed studies and mainstream topics.

This disconnect between policy and real research needs has created an environment where unconventional ideas are ignored, independent researchers are underfunded, and emerging fields struggle to gain recognition. Instead of encouraging pioneering scientific advancements, current policies tend to reinforce existing academic hierarchies, making it nearly impossible for unaffiliated scholars to contribute meaningfully.

This section explores how academic publishing policies, government funding priorities, and policy decisions neglect the real needs of researchers, particularly those working outside institutional frameworks.

6.3.1 How Academic Publishing Policies Favor Mainstream Topics Over Unconventional Ideas

For a researcher to gain academic recognition and funding, their work must be published in respected journals and conferences. However, academic publishing favors research in mainstream, well-established fields, making it difficult for researchers exploring unconventional, interdisciplinary, or emerging topics to get published.

1. Preference for Established Research Over Innovative Work

- Many journals and funding agencies favor "safe" research—topics that build upon existing theories and popular scientific trends rather than exploring new and radical ideas.
- Studies that challenge conventional scientific thought or propose paradigm shifts often face higher scrutiny and rejection rates.

- This prevents disruptive scientific progress, as researchers are pressured to align with existing academic norms rather than take bold risks.

2. Resistance to Cross-Disciplinary and Independent Research

- Many high-impact journals expect research to fit into well-defined academic categories, making it difficult for cross-disciplinary studies to get published.
- Independent researchers often explore unique perspectives, combining elements of physics, AI, biology, or philosophy, but their work does not fit into traditional research silos, leading to rejection.
- The peer review process is dominated by experts in specific fields, who may resist ideas that challenge existing academic boundaries.

3. Slow Adoption of New Research Areas

- Many groundbreaking fields (e.g., quantum biology, AI ethics, neurophysics) took decades to gain recognition because academic journals delayed accepting research in these areas.
- Emerging fields often gain credibility only when backed by elite universities, leaving independent researchers struggling to publish until the field becomes mainstream.
- This system delays scientific progress and discourages independent innovation.

By favoring established research over disruptive innovation, academic publishing restricts the growth of knowledge and excludes independent scholars who challenge conventional wisdom.

6.3.2 Why Government Funding Bodies Prioritize Institutional Research Over Independent Projects

Most scientific breakthroughs require funding, but government research grants and funding bodies overwhelmingly prioritize institutional researchers, leaving independent scholars without financial support.

1. Funding Eligibility Criteria Exclude Independent Researchers

- Most government research grants require an institutional affiliation, effectively disqualifying independent researchers from applying.
- Funding applications often ask for:

- Institutional recommendations.
- Research lab affiliations.
- Institutional approval for financial oversight.

- Without these credentials, independent researchers cannot secure funding, even if their research is scientifically valid and socially impactful.

2. Preference for Large-Scale Institutional Projects Over Individual Research

- Government grants are often awarded to high-budget, large-scale projects involving multiple university departments, while small independent projects struggle to receive attention.
- Even when independent researchers propose low-cost, high-impact projects, funding agencies prioritize institutional teams with extensive proposals and bureaucratic support.
- As a result, many innovative projects fail to secure funding simply because they do not come from recognized institutions.

3. Institutional Favoritism in Research Grant Approvals

- In many cases, research funding is allocated based on institutional reputation rather than the merit of the research proposal.
- Universities with strong government connections and political influence secure larger portions of national research grants, while independent researchers are ignored.
- Funding is often directed towards projects aligned with government priorities, discouraging independent exploration of socially or scientifically relevant topics.

This bias in funding distribution makes it nearly impossible for independent researchers to sustain their work, forcing them to either self-finance their projects or abandon research altogether.

6.3.3 How Policy Decisions Ignore the Needs of Independent Researchers in Emerging Fields

While policymakers discuss scientific progress and innovation, their actual policies often fail to support independent researchers—especially those working in emerging and unconventional fields.

1. The Absence of Support Systems for Independent Researchers

- Unlike countries like Germany, the US, and China, India and many developing nations lack structured programs for independent research funding.
- Independent researchers have no access to subsidized labs, library resources, or government-backed research fellowships, making it difficult for them to contribute to national scientific progress.
- Many countries do not even recognize independent

researchers in national R&D policy frameworks, treating them as outsiders rather than contributors to scientific development.

2. Emerging Research Fields Receive Little Government Attention

- Many pioneering scientific fields—such as biotech ethics, decentralized AI, nanorobotics, and climate engineering—are underfunded or ignored by government funding agencies.
- Since these fields are still evolving, policymakers hesitate to allocate resources, forcing independent researchers to struggle without institutional support.
- By the time government agencies recognize the importance of an emerging field, institutional researchers dominate the space, leaving independent scholars without credit or funding.

3. The Lack of Policy Reforms for Research Autonomy

- Many countries have rigid academic policies that make it difficult for independent researchers to contribute to national research initiatives.
- Policies should be updated to:

- Allow independent researchers to apply for national grants and fellowships.

- Provide research facilities and lab access to unaffiliated scholars.

- Recognize independent contributions in national research metrics and reports.

- Without these changes, independent researchers will continue to be excluded from academic and scientific policymaking.

By failing to address the needs of independent researchers, policy decisions slow down scientific progress, discourage innovation, and reinforce academic elitism.

Conclusion

The disconnect between policy and real research needs has created a biased system where independent researchers:

- Struggle to publish unconventional work due to academic conservatism.
- Are systematically excluded from government funding opportunities.
- Receive no policy support for contributing to emerging research fields.

If governments and academic institutions truly want to advance scientific discovery, they must:

- Revise publishing policies to encourage innovative and independent research.

- Remove institutional barriers in research funding applications.

- Create government-backed fellowships and research centers for independent scholars.

Without these policy changes, academia will continue to be controlled by institutional elites, preventing brilliant independent researchers from making transformative contributions to science and society.

6.4 The Challenges of Peer Review: Institutional Favoritism and Gatekeeping

The peer review process is intended to be a rigorous, objective, and fair evaluation system that ensures the quality and credibility of academic research. However, in reality, peer review often operates as a gatekeeping mechanism that favors institutional researchers while excluding independent scholars from publishing in high-impact journals.

Rather than being an impartial assessment of research quality, peer review is often influenced by institutional affiliations, academic networking, and hierarchical biases. This system makes it significantly harder for independent researchers to gain credibility, as their work is scrutinized more harshly than institution-backed studies.

This section explores how institutional favoritism, credibility challenges, and academic networking biases influence peer review, preventing merit-based evaluation of independent research.

6.4.1 How Peer Reviewers Often Favor Institution-Backed Research

Peer review is meant to evaluate the content of research, not the researcher's background. However, many reviewers—whether consciously or unconsciously—show bias toward research affiliated with prestigious institutions.

1. The Unwritten Bias Toward Elite Universities

- Reviewers, particularly those from top-tier institutions, tend to trust and favor submissions from well-known universities like Harvard, MIT, Oxford, or IITs over independent or unknown researchers.
- A submission from a Cambridge University scientist is often viewed as more credible than a similar submission from an independent researcher, even if the latter's work is of equal or

better quality.

- This leads to a self-reinforcing cycle, where institution-backed research gets easier acceptance, while independent research is frequently dismissed.

2. The "Institutional Safety Net" in Peer Review

- Researchers from elite institutions often receive the benefit of the doubt—if there are minor flaws in their papers, reviewers suggest improvements instead of rejecting the paper.
- Independent researchers, however, face immediate rejection for similar flaws because they lack institutional backing to reinforce their credibility.
- Some reviewers assume that independent researchers lack access to peer guidance, leading them to view unaffiliated work as less trustworthy or rigorous.

3. Institutional Affiliation Influences the Perception of Research Validity

- Many reviewers rely on familiar academic networks, leading to a subconscious bias where research from known institutions is considered "more reliable" than unaffiliated research.
- Studies have shown that when reviewers are unaware of an author's affiliation, research from independent scholars is evaluated more fairly than when affiliation is disclosed.

This suggests that institutional favoritism is a systemic issue in peer review, impacting acceptance rates and limiting opportunities for independent researchers.

6.4.2 Why Independent Researchers Struggle with Credibility in the Peer Review Process

Credibility in academia is built over time through published work, citations, and institutional affiliations. However, independent researchers lack institutional backing, making it difficult for them to establish academic legitimacy.

1. The Lack of an Institutional Identity in Academic Publishing

- Many journal reviewers look for institutional credentials as a mark of credibility.
- Independent researchers do not have access to university endorsement letters, reference committees, or faculty advisors, making them easier to dismiss in the peer review process.
- Some journals even outright reject submissions from unaffiliated researchers or make affiliation a mandatory requirement for publication.

2. Difficulty in Passing the "First Impression Test"

- Before even reading a paper in depth, many reviewers scan the author's credentials and affiliation, forming an initial judgment about the research's reliability.
- Without a recognizable institution attached to their name, independent researchers start at a disadvantage in the review process.
- Even if the research is strong, reviewers may assume it lacks adequate mentorship, lab verification, or academic guidance, leading to higher rejection rates.

3. Independent Researchers Receive Harsher Scrutiny

- Research from recognized institutions is often reviewed more leniently, while independent research is subjected to excessive questioning.
- Even well-written papers with solid methodologies and strong

findings are sometimes rejected because reviewers doubt the research environment in which they were conducted.

- Without institutional credibility to reinforce their claims, independent researchers must go to extreme lengths to justify their methodologies and results.

6.4.3 The Influence of Academic Networks in Determining Publication Success

Beyond institutional bias, academic networking plays a crucial role in determining whose research gets published. Researchers who are well-connected in academic circles—especially those from elite institutions—have an easier time getting their work reviewed, accepted, and cited.

1. The Power of Academic Networking in Peer Review

- Many high-profile researchers serve as peer reviewers and editorial board members, and they tend to favor work from scholars within their academic circles.
- Institutional researchers often attend conferences, collaborate on projects, and interact with reviewers in academic settings, giving them a strategic advantage in the peer review process.
- Independent researchers, without these academic connections, struggle to get their work noticed and fairly evaluated.

2. The Hidden Impact of Citation and Reference Bias

- Journals and reviewers often expect authors to cite research from established academics in their field.
- If an independent researcher fails to reference prominent institutional studies, their work may be considered incomplete or less credible.

- This expectation reinforces the dominance of institutional research, as reviewers often prioritize research that aligns with widely accepted academic perspectives.

3. The Advantage of Institutional Collaboration in Peer Review

- Many independent researchers struggle to publish because they lack co-authors from prestigious institutions.
- Papers co-authored with senior academics from top universities often have higher chances of acceptance, even if the independent researcher was the primary contributor.
- This forces many independent scholars to seek institutional affiliations or collaborations, not for scientific merit but simply to improve publication success.

Conclusion

The peer review system, which should be an unbiased process, is heavily influenced by institutional favoritism, networking, and systemic biases. This creates significant barriers for independent researchers, making it difficult for them to gain credibility, publish their work, and contribute to academic discourse.

Key challenges faced by independent researchers include:

- Bias toward institution-backed research, leading to higher rejection rates for unaffiliated scholars.
- A lack of institutional credibility, making it harder for independent researchers to pass the "first impression test" in peer review.
- The dominance of academic networking, where well-connected researchers from elite institutions receive preferential treatment.

To create a fairer peer review system, academic publishing must:

- Adopt double-blind review processes to eliminate institutional bias.
- Encourage journals to accept high-quality research from independent scholars, regardless of affiliation.
- Increase transparency in the peer review process to prevent favoritism and institutional gatekeeping.

Without these reforms, independent researchers will continue to be excluded from mainstream academic publishing, reinforcing the monopoly of institutional research and limiting diverse contributions to global knowledge.

6.5 Predatory Journals and the Exploitation of Independent Researchers

For independent researchers struggling to publish in reputable journals, the academic publishing landscape presents an additional challenge—predatory journals. These journals operate without proper peer review, charge exorbitant publication fees, and exploit unaffiliated scholars who lack access to institutional resources.

While institutional researchers receive mentorship, funding, and access to legitimate publishing platforms, independent scholars often find themselves misled by deceptive journals that promise quick publication but ultimately damage their academic credibility. The combination of lack of awareness, financial pressure, and desperation to publish forces many independent researchers into the trap of pay-to-publish journals that offer little academic recognition.

This section examines how predatory journals take advantage of independent researchers, why many scholars unknowingly publish in them, and how these exploitative practices impact the academic careers of unaffiliated researchers.

6.5.1 How Predatory Journals Take Advantage of Unaffiliated Researchers

Predatory journals exist solely to make money, targeting researchers who are desperate to publish quickly and establish credibility. They exploit independent researchers by offering:

1. Fast-Track Publications Without Real Peer Review

- Many independent scholars face high rejection rates from reputable journals, leading them to seek faster publication routes.
- Predatory journals promise acceptance within days or weeks, bypassing the traditional peer review process.
- They accept almost any submission, regardless of quality, which devalues the research and damages the author's reputation.

2. High Publication Fees Without Academic Value

- Reputable journals charge processing fees but ensure rigorous peer review and indexing in academic databases.
- Predatory journals charge fees ranging from $500 to $5,000 per article but provide no real academic validation.
- Since independent researchers lack institutional funding, they often pay these fees out of pocket, leading to financial exploitation.

3. Fake Impact Factors and Misleading Indexing Claims

- Many predatory journals falsely claim to be indexed in Scopus, Web of Science, or PubMed, luring researchers into believing their work will gain credibility.
- They often advertise fake impact factors or cite meaningless metrics to create an illusion of legitimacy.
- Unaffiliated researchers, without proper guidance from academic mentors, fall victim to these claims and

unknowingly publish in non-reputable outlets.

Once a researcher publishes in a predatory journal, their work is often ignored by the academic community, making it difficult for them to gain recognition in legitimate research circles.

6.5.2 The Lack of Awareness About Legitimate vs. Exploitative Publishing Platforms

One of the biggest reasons independent researchers fall into the predatory journal trap is the lack of awareness about legitimate publishing platforms. Without institutional support, they struggle to differentiate between credible and deceptive journals.

1. Difficulty in Identifying Reputable Journals

- University researchers have access to research librarians and faculty advisors who guide them in selecting reputable journals.
- Independent researchers, however, must navigate the publishing world alone, making them more vulnerable to misleading journal websites and fake editorial boards.
- Many predatory journals imitate the names of established journals, tricking scholars into submitting their work.

2. The Absence of Training in Journal Selection

- Many independent researchers lack formal training on how to evaluate journal credibility.
- Without academic mentorship, they cannot distinguish between open-access journals with rigorous peer review (like PLOS ONE) and fraudulent journals that accept any submission.
- Some predatory publishers misuse open-access models, leading independent scholars to confuse pay-to-publish scams

with legitimate open-access initiatives.

3. Over-Reliance on "Calls for Papers" Emails and Mass Invitations

- Predatory journals aggressively send email invitations promising quick publication, which many independent researchers mistake for genuine academic interest.
- Since unaffiliated scholars are often excluded from mainstream publishing networks, they feel validated when approached by such journals, not realizing it is a scam.
- Many researchers submit their work out of urgency, only to later discover that the journal lacks academic credibility and recognition.

The lack of formal academic guidance and training puts independent researchers at a high risk of falling into predatory publishing traps.

6.5.3 How Independent Scholars Fall Into the Trap of Pay-to-Publish Journals

The pressure to publish combined with limited access to reputable journals forces many independent researchers into pay-to-publish schemes, leading to long-term academic consequences.

1. Desperation to Publish for Academic Recognition

Many independent scholars feel pressured to establish their academic presence, especially if they:

- Are applying for funding opportunities.
- Want to transition into academic roles.
- Need to build credibility for consultancy work.

Since mainstream journals reject a high percentage of independent submissions, predatory journals offer them an easy alternative—one that often comes with long-term consequences.

2. The Financial Burden of Pay-to-Publish Models

- Without institutional grants or university funding, independent researchers must personally pay hefty publication fees.
- Some predatory publishers operate tiered pricing models, charging researchers extra fees for so-called "impact factor boosts" or "faster processing."
- Many scholars end up spending large amounts of money on low-quality journals, depleting their financial resources while gaining no real academic recognition.

3. The Negative Impact on Career Growth and Academic Credibility

- Once a researcher publishes in a predatory journal, their work is often disregarded by academic institutions and funding agencies.
- Some universities even blacklist researchers who have published in predatory journals, preventing them from applying for faculty positions or research grants.
- Even if the research itself is valuable, its association with a low-quality journal makes it less likely to be cited or taken seriously.

Many independent scholars realize their mistake only after publication, by which time their academic reputation is already affected.

Conclusion

Predatory journals exploit independent researchers, taking advantage of their:

- Desperation to publish and gain credibility.
- Lack of awareness about journal legitimacy.
- Limited financial resources, forcing them into pay-to-publish schemes.

To prevent exploitation, independent researchers must:

- Verify journal credibility using databases like Scopus, Web of Science, and DOAJ.
- Consult platforms like Think. Check. Submit. to evaluate journals before submitting.
- Avoid responding to unsolicited "Calls for Papers" from unknown publishers.

Use open-access preprint servers (like arXiv, SSRN, bioRxiv) to gain visibility before formal publication.

Meanwhile, academic policymakers must:

- Raise awareness about predatory publishing through workshops and online resources.
- Create funding schemes for independent researchers to publish in reputable journals.
- Encourage more rigorous peer review models in open-access publishing.

Without these interventions, independent researchers will continue to fall prey to predatory publishing scams, further widening the gap between institution-backed research and unaffiliated scholarly contributions.

6.6 Citation Bias and the Struggle for Recognition

Citations play a crucial role in academic recognition, influencing a researcher's credibility, funding opportunities, and career advancement. A well-cited paper gains visibility, impact, and acknowledgment within the academic community, while research with low citations struggles to be taken seriously.

However, citation patterns are not always based on merit—they are heavily influenced by institutional affiliations, academic networking, and journal prestige. Independent researchers, despite producing high-quality work, often receive fewer citations because their research is overlooked or excluded from mainstream academic discussions.

This section explores how citation bias impacts independent researchers, focusing on the dominance of elite universities, institutional citation networks, and the consequences of low citation counts on research credibility and funding.

6.6.1 Why Independent Research Receives Fewer Citations Than Institutional Studies

In theory, citations should reflect the quality and impact of research, but in reality, they are often skewed by academic hierarchies and institutional privilege. Independent researchers face multiple barriers that prevent their work from being widely cited and recognized.

1. Institutional Branding Increases Citation Visibility

- Researchers from Harvard, MIT, Stanford, and other elite institutions have a higher chance of being cited, even if their work is similar to—or less impactful than—independent research.
- Papers from renowned universities are automatically considered more credible, leading scholars to prefer citing

institution-backed studies over independent ones.

- Journals and reviewers often recommend citing papers from well-known academics, reinforcing citation bias.

2. Independent Researchers Lack Institutional Amplification

- University researchers benefit from built-in citation networks, where faculty members, research groups, and students cite each other's work in publications.
- Independent scholars lack institutional affiliations, making it harder for them to get referenced in high-impact papers.
- Many independent studies are not indexed in mainstream citation databases, further reducing their visibility.

3. Limited Access to High-Impact Journals Reduces Citations

- Papers published in high-impact journals (e.g., Nature, Science, IEEE Transactions) receive more citations because they are widely read and referenced.
- Independent researchers struggle to publish in these journals, meaning their work remains hidden in lower-tier or niche publications.
- Since many academics only reference papers from top-tier journals, independent researchers receive fewer citations even if their work is valuable.

These systemic disadvantages mean that independent research is often ignored or under-cited, leading to a loss of academic recognition.

6.6.2 How Elite Universities Dominate Citation Networks

Citation networks are heavily influenced by institutional hierarchies, where researchers from top universities dominate academic references and publication cycles.

1. Self-Citation and Institutional Citation Networks

- Many university researchers cite their colleagues, mentors, and faculty members, creating closed-loop citation networks that exclude independent scholars.
- Some research groups systematically reference their previous work to boost citation metrics, while independent researchers have no such support system.
- Well-known academics in elite institutions collaborate on multi-author papers, leading to higher citation counts across institutional teams.

2. The Influence of Journal Prestige on Citation Patterns

- Journals affiliated with elite universities receive more citations because they are considered more credible and widely read.
- Independent researchers who publish in lesser-known or open-access journals often get ignored in citation databases, even if their work is groundbreaking.
- Many high-impact papers deliberately reference studies from top-tier institutions, reinforcing institutional dominance.

3. Institutional Access to Research Conferences and Networks

- Top university researchers attend global conferences, give keynote speeches, and participate in high-profile panels, where their work is discussed, referenced, and cited.
- Independent researchers often cannot afford to attend these events, meaning their work remains isolated from mainstream academic discussions.
- Since research ideas and citations often emerge from conference discussions, independent scholars are left out of the citation loop.

This academic hierarchy ensures that elite universities continue to dominate citation networks, making it harder for independent researchers to gain recognition.

6.6.3 The Impact of Low Citation Counts on Funding and Research Credibility

A paper's citation count is often used as a measure of impact, influencing funding decisions, research grants, and academic career prospects. However, for independent researchers, low citation counts create a vicious cycle of academic exclusion.

1. Low Citation Counts Lead to Funding Rejections

- Many funding agencies evaluate researchers based on their citation metrics, assuming that highly cited work is more impactful.
- Independent researchers, with fewer citations due to systemic bias, struggle to convince grant committees of their research's value.
- Even when independent research is scientifically sound, its lack of citations makes it less competitive for funding opportunities.

2. Research Credibility Is Tied to Citation Metrics

- In academia, research with high citations is considered authoritative, while low-citation work is seen as less influential.
- This leads to a credibility gap, where independent researchers are dismissed, not because of research quality, but due to citation statistics.
- Many universities and journals use h-index scores (a measure of citations per author), putting independent researchers at a severe disadvantage.

3. Career Limitations for Independent Scholars

- Many academic positions require a strong citation record, making it difficult for independent researchers to secure faculty roles or research fellowships.
- Citation bias creates a self-reinforcing system, where independent researchers struggle to get recognized, publish in top journals, or secure academic positions.
- Even when an independent researcher produces high-quality work, their lack of citations results in continued professional marginalization.

These challenges demonstrate that low citation counts are not just a statistical issue—they actively limit the careers, funding, and credibility of independent scholars.

Conclusion

Citation bias is one of the biggest barriers to independent research recognition. The academic system favors elite universities and institutional networks, making it difficult for unaffiliated researchers to:

- Get cited, even when their research is valuable.
- Enter academic discussions dominated by institutional researchers.
- Secure funding, career opportunities, and professional credibility.

To address citation bias, academia must:

- Encourage more open-access citations, allowing independent researchers to gain visibility.
- Promote citation diversity, discouraging self-referential citation networks in elite universities.

- Develop funding programs that evaluate research based on merit, not citation metrics alone.

Without these changes, independent researchers will remain locked out of academic recognition, reinforcing a system where institutional prestige outweighs scientific contribution.

6.7 Overcoming Barriers: Solutions for Independent Researchers

Despite the numerous challenges faced by independent researchers in publishing, recognition, and funding, there are several alternative pathways and strategies that can help them gain credibility, visibility, and access to global research communities.

While traditional academic systems continue to favor institutional research, independent scholars can leverage open-access platforms, global networks, preprint servers, industry collaborations, and digital tools to bypass institutional gatekeeping and establish their research impact.

This section explores practical solutions that independent researchers can adopt to overcome publication biases, build research credibility, and integrate into global academic conversations.

6.7.1 Leveraging Open-Access Journals and Alternative Publishing Platforms

One of the biggest obstacles for independent researchers is getting published in high-impact journals, which often have institutional biases and high publication fees. However, open-access journals and alternative publishing platforms provide new opportunities for independent scholars to publish and share their work globally.

1. Publishing in Reputable Open-Access Journals

Unlike traditional journals that restrict access behind paywalls, open-access journals make research freely available, increasing visibility and citations. Well-established open-access journals include:

- PLOS ONE (multidisciplinary, rigorous peer review).
- Frontiers (various research fields, open-peer review).
- MDPI Journals (fast publication, indexed in major databases).

Open-access platforms allow independent researchers to reach a wider audience, including industry professionals, policymakers, and the public.

2. Using Alternative Research Platforms for Visibility

Platforms like ResearchGate, Academia.edu, and SSRN allow researchers to:

- Share preprints and published work for wider dissemination.
- Engage in discussions with researchers from different fields and backgrounds.
- Increase visibility outside institutionally controlled journal databases.

These platforms help independent researchers bypass traditional academic hierarchies and build their own credibility.

By strategically using open-access and alternative publishing routes, independent researchers can establish a strong academic presence without relying on traditional institutional backing.

6.7.2 Building Global Networks and Online Research Communities

Independent researchers often struggle with academic isolation, as they lack access to faculty mentors, institutional collaborations, and peer groups. However, online networks and global research communities can help them connect with like-minded scholars, share ideas, and receive constructive feedback.

1. Joining Online Research Communities

Many international organizations and platforms support independent research collaborations, such as:

- The Global Young Academy (GYA) – connects early-career researchers worldwide.
- The Open Science Framework (OSF) – provides tools for research collaboration and data sharing.
- Citizen Science Networks – enable researchers to participate in global scientific projects.

Engaging in these communities allows independent researchers to access mentorship, collaborate on interdisciplinary projects, and gain international exposure.

2. Participating in Virtual Conferences and Webinars

Many global research conferences now offer virtual participation, reducing the cost barrier for independent scholars. Attending and presenting at online conferences, webinars, and workshops helps researchers:

- Build their academic reputation.
- Receive feedback from peers and experts.
- Network with industry professionals and research funders.

By leveraging online research communities and virtual conferences, independent scholars can expand their academic network beyond traditional institutions.

6.7.3 Using Preprint Servers to Gain Visibility Before Peer Review

One major challenge independent researchers face is the lengthy peer-review process, which can take months or even years before research is published. Preprint servers provide a solution by allowing researchers to share their work publicly before formal peer review, ensuring early visibility and feedback.

1. What Are Preprint Servers?

Preprint servers allow researchers to upload manuscripts for public access before submission to peer-reviewed journals.

They provide a timestamped record of research, ensuring that independent scholars receive credit for their discoveries.

Leading preprint servers include:

- arXiv.org (Physics, Computer Science, Math).
- bioRxiv (Biology and Life Sciences).
- medRxiv (Medical and Health Research).
- PsyArXiv (Psychology and Behavioral Sciences).

2. Benefits of Using Preprints for Independent Researchers

- **Faster dissemination** – Work is available immediately instead of waiting for peer review.
- **More citations** – Studies show that preprints receive higher citation rates over time.
- **Increased collaboration** – Researchers worldwide can see and discuss the work, leading to potential research partnerships.
- **Alternative validation** – Even if a paper is not accepted by traditional journals, it can still gain credibility through community engagement.

By using preprint servers, independent researchers can gain early recognition and feedback, increasing their chances of getting noticed by journals and academic peers.

6.7.4 Collaborating with Industry and Private Organizations for Research Support

Since independent researchers lack institutional funding, collaboration with private industry, tech companies, and research-focused organizations can provide financial support, data access, and research facilities.

1. Partnering with Industry for Research Funding

- Many corporations and R&D departments fund research in AI, biotechnology, engineering, and applied sciences.
- Companies like Google, Microsoft, Tesla, and pharmaceutical firms sponsor independent research in emerging fields.
- Platforms such as MIT Solve, Gates Foundation, and XPRIZE offer funding for independent innovators working on global challenges.

2. Accessing Industry Labs and Data Resources

Some companies and private labs allow independent researchers to use their facilities for experimentation. Collaborating with tech firms, research startups, or government think tanks can provide access to:

- Advanced computing resources (e.g., AI research in cloud computing).
- Medical and biotech databases (e.g., genomics and drug research).
- Financial and economic datasets (e.g., behavioral research in market trends).

By collaborating with industry, independent researchers can overcome resource limitations and conduct high-quality research outside of traditional academic institutions.

6.7.5 Developing Independent Research Collectives for Credibility and Funding

Independent researchers often struggle with individual credibility, but forming collective research groups can improve their standing in the academic community.

1. Establishing Independent Research Associations

Researchers can create small, self-funded research collectives under a formal nonprofit or research organization structure. Examples of successful non-institutional research groups include:

- The Santa Fe Institute (Complex Systems Research).
- The Ronin Institute (Independent Scholars Network).

These collectives provide collaborative publishing opportunities, shared funding resources, and institutional legitimacy.

2. Crowdfunding and Open Science Grants

- Independent scholars can raise funds through platforms like Kickstarter, GoFundMe, or Patreon to support their research.
- Open Science initiatives such as Mozilla Open Science and OpenAI Grants provide funding for non-traditional research projects.
- By organizing research collectives and securing alternative funding sources, independent researchers can gain recognition and publish under credible research institutions.

6.7.6 Utilizing AI and Digital Tools to Enhance Research Visibility and Networking

AI-powered platforms and digital tools can help independent researchers increase their visibility, automate research workflows, and expand their academic reach.

1. AI-Based Research Assistance

Tools like ChatGPT, Elicit, and Scite help researchers:

- Analyze vast amounts of literature quickly.
- Identify relevant citations and sources.
- Summarize key findings for faster research processing.

2. Enhancing Research Discoverability with AI

AI-driven platforms such as Semantic Scholar and Connected Papers help researchers:

- Find new citation opportunities.
- Connect with relevant academic discussions.
- Increase their work's discoverability in academic circles.

By leveraging AI and digital tools, independent researchers can enhance their visibility and engagement with the global research community.

Conclusion

While independent researchers face institutional biases, they can overcome these barriers by:

- Leveraging open-access platforms and preprints for wider visibility.
- Building global research networks for credibility and support.
- Collaborating with industry and forming independent research collectives.
- Utilizing AI tools to optimize research workflows and discoverability.

By taking strategic steps, independent scholars can establish themselves in the academic world, proving that research excellence is not limited to institutional affiliation.

7. The Need for Policy Reform: What India Must Do

The previous chapters highlighted the systemic barriers faced by independent researchers in India—from institutional bias in publishing and funding restrictions to lack of lab access and recognition in citation networks. These challenges are not just individual struggles but reflect a deeper structural problem in India's research ecosystem, where policy frameworks overwhelmingly favor institutional research while ignoring independent scholars.

If India wants to foster a truly innovative and knowledge-driven economy, it must rethink its approach to research funding, infrastructure, and recognition. Independent researchers play a crucial role in global innovation, often driving breakthrough discoveries outside traditional academic settings. However, without financial support, access to facilities, and recognition in national research policies, their potential remains largely untapped.

This chapter explores key policy reforms that India must implement to create a more inclusive, supportive, and innovation-driven research ecosystem. From establishing direct funding mechanisms and providing lab access to encouraging corporate R&D investments and adopting global best practices, these reforms can empower independent researchers and position India as a global leader in scientific progress.

7.1 Establishing Direct Funding Mechanisms for Independent Researchers

Research funding is the lifeline of scientific progress, yet India's current funding structure is heavily skewed toward institutional researchers, leaving independent scholars without financial support or access to critical resources. Government research grants, fellowships, and subsidies are typically restricted to university-affiliated scientists, creating an exclusionary system where independent researchers are forced to self-fund their projects or abandon their work altogether.

If India wants to maximize its innovation potential, it must establish direct funding mechanisms for independent researchers by:

- Creating government grants specifically tailored for non-institutional scholars.
- Reducing bureaucratic obstacles in research funding applications.
- Providing tax benefits and financial incentives for self-funded research initiatives.

This section explores how India can implement these funding solutions to empower independent researchers and ensure that scientific progress is not restricted to academic institutions alone.

7.1.1 Government Research Grants Tailored for Independent Scholars

In most developed countries, independent researchers have direct access to government-funded grants and fellowships, allowing them to pursue groundbreaking research outside of traditional academic institutions. However, in India, almost all government research funding is allocated to universities, IITs, IISc, and CSIR labs, leaving independent researchers with little to no financial support.

To correct this imbalance, India must establish dedicated research grant programs for independent scholars, modeled after successful global initiatives.

1. Creating a National Independent Research Grant Program

The Indian government should establish a National Independent Research Grant (NIRG), similar to:

- **The European Research Council (ERC) Grants** – which fund individual researchers, not just institutions.
- **The National Science Foundation (NSF) in the US** – which supports independent and early-career researchers.
- **China's National Natural Science Foundation (NSFC)** – which provides grants to independent innovators.

The NIRG should provide direct funding to independent researchers, allowing them to apply for research grants without needing institutional affiliation.

It should include funding categories for:

- Early-career independent researchers (who lack university support).
- Experienced scientists who wish to work independently outside academia.
- Innovators working on interdisciplinary and emerging fields that are not widely recognized by traditional funding agencies.

2. Offering Seed Grants and Project-Based Funding for Independent Researchers

- In addition to large grants, India should introduce seed funding programs that offer ₹10–50 lakh (₹1–5 million) for short-term independent research projects.

- These grants should be project-based, allowing researchers to apply for funding based on the merit and potential impact of their work, rather than their institutional affiliation.
- Successful independent researchers should have access to follow-up grants based on the progress of their initial projects.

By establishing government-backed research grants specifically for independent scholars, India can unlock the full potential of non-institutional research and encourage more scientists to pursue independent innovation.

7.1.2 Reducing Bureaucratic Barriers in Funding Applications

Even when funding is available, India's research grant application process is complex, bureaucratic, and often inaccessible to independent scholars. Institutional researchers benefit from university administrative support, whereas independent researchers must navigate the system alone, facing excessive paperwork, delays, and rejection.

To ensure that independent researchers can actually access funding, India must simplify the grant application process and remove unnecessary administrative barriers.

1. Creating a Streamlined Online Portal for Research Funding Applications

- India should develop a single-window digital portal where independent researchers can apply for grants, track their applications, and receive funding without bureaucratic delays.
- The system should: Eliminate redundant paperwork by allowing researchers to submit proposals online. Provide transparent timelines for application review and funding disbursement. Offer direct communication channels for applicants to receive feedback.

- This will ensure faster processing of grant applications and reduce the administrative burden on independent researchers.

2. Removing Institutional Endorsement Requirements

- Many Indian research grants require applicants to have a letter of recommendation from an institution or university faculty—a requirement that automatically excludes independent researchers.
- This policy should be revised to allow self-sponsored applications, ensuring that independent scholars do not need institutional approval to access funding.

3. Establishing a Fast-Track Review Process for Independent Researchers

- Since independent researchers lack the institutional support that speeds up university-backed applications, India must establish: A special review committee for independent research proposals. Shorter turnaround times for application review and grant disbursement. A resubmission process with detailed feedback for rejected applications.
- These measures will increase accessibility and efficiency, ensuring that independent researchers receive funding without unnecessary delays.
- By simplifying the research funding process, India can make government grants more accessible to independent scholars and ensure that valuable research projects do not get lost in bureaucratic red tape.

7.1.3 Introducing Tax Benefits and Financial Incentives for Self-Funded Researchers

Since independent researchers lack institutional funding, many are forced to pay for research expenses out of their own pockets. This financial burden often leads to abandoned projects, incomplete studies, and wasted research potential.

To support self-funded researchers, the Indian government must introduce tax incentives and financial relief measures that reduce the economic strain on independent scholars.

1. Providing Tax Deductions for Self-Funded Research Expenses

Independent researchers should be allowed to deduct research-related expenses from their taxable income, including:

- Laboratory and equipment costs.
- Journal publication fees.
- Data access and software purchases.
- Travel expenses for research-related conferences.

This policy would function similarly to tax deductions for startup investments, encouraging more researchers to self-fund their projects without financial hardship.

2. Offering Research Fellowships with Monthly Stipends for Independent Scholars

- In addition to grants, India should introduce monthly research stipends for independent scholars, similar to:
- The Humboldt Fellowship (Germany) – which provides financial support to non-institutional researchers.
- The Marie Curie Fellowship (EU) – which funds independent researchers to work freely in their field.

A ₹50,000–₹1,00,000 monthly stipend for independent scholars would:

- Help cover living and research expenses.
- Enable researchers to dedicate time to full-time independent work.
- Prevent the loss of promising researchers due to financial constraints.

3. Creating a Research Investment Fund for Crowdsourced and Publicly-Funded Research

India should establish a National Research Investment Fund, allowing citizens and corporations to donate toward independent research with tax-exempt benefits. This would encourage public-private participation in funding innovation, similar to:

- The Wellcome Trust (UK) – a charity-funded research institution.
- The Gates Foundation (US) – which funds non-institutional global research.

This model would democratize research funding, allowing independent scholars to receive support directly from the public and industry.

By introducing tax benefits, stipends, and crowdfunding options, India can reduce the financial burden on independent researchers, making it easier for them to pursue high-impact projects without institutional dependence.

Conclusion

For India to fully utilize the potential of its independent researchers, the government must:

- Establish direct research grants for non-institutional scholars.
- Reduce bureaucratic barriers in funding applications.
- Introduce tax incentives and financial relief for self-funded research.

These policies will create a more inclusive and dynamic research ecosystem, ensuring that scientific discovery is driven by innovation, not just institutional privilege.

7.2 Providing Lab Access, Research Fellowships, and Mentorship Programs

One of the biggest challenges for independent researchers in India is the lack of access to essential research infrastructure, fellowships, and academic mentorship. Unlike university-affiliated scholars who have access to state-of-the-art labs, research grants, and faculty guidance, independent researchers struggle to find workspaces, experimental facilities, and expert advice.

For India to fully utilize the potential of its independent researchers, it must introduce policy reforms that open up lab access, create dedicated research fellowships, and establish structured mentorship programs. These changes will help bridge the gap between institutional and independent research, ensuring that all scientists, regardless of affiliation, have equal opportunities to contribute to innovation and knowledge creation.

7.2.1 Opening University and Government Research Labs to Unaffiliated Scholars

One of the biggest barriers for independent researchers is the exclusion from university and government research labs. Most labs in IITs, IISc, CSIR institutes, and public universities are restricted to institutional researchers, making it impossible for independent scholars to conduct high-level experiments, access scientific equipment, or collaborate on large-scale research projects.

Challenges in Lab Access for Independent Researchers

- Universities do not allow independent researchers to use lab facilities unless they are affiliated with the institution.
- Government-funded research centers restrict access to

officially enrolled students and faculty members.

- Private research labs charge extremely high fees for independent researchers to conduct experiments.
- Lack of shared research spaces makes it difficult for unaffiliated scholars to access essential scientific tools.

To solve this problem, India must open up its government and university research labs to independent scholars.

Proposed Solutions: Unlocking Lab Access for Independent Researchers

1. Establishing a National Lab Sharing Program for Independent Researchers

- The Indian government should introduce a National Lab Sharing Program, allowing independent researchers to apply for temporary lab access in government-funded institutions.
- This program can be modeled after Germany's Max Planck Institute and the US National Laboratory User Facility Program, which allow external researchers to use their facilities.
- Researchers should be able to book lab time in public institutions based on merit and project feasibility.

2. Providing Subsidized or Free Lab Access for Independent Scholars

- Government research labs should reserve a percentage of lab hours exclusively for independent researchers.
- Institutions should provide subsidized or free access to equipment, chemicals, and technology tools for unaffiliated scholars.
- State-funded research institutions like IISc and CSIR labs should allow independent researchers to apply for shared

workspace access.

3. Creating a Centralized Lab Access Portal

- A national digital portal should be created where independent researchers can apply for access to government and university labs.
- The portal should: List all available lab facilities across the country. Allow independent researchers to book lab slots for specific projects. Enable collaboration between independent scholars and institutional researchers.
- By providing lab access to independent researchers, India can ensure that scientific progress is not restricted to institutional boundaries.

7.2.2 Creating National-Level Fellowships Specifically for Independent Researchers

Independent researchers lack access to structured funding programs, which forces many of them to self-fund their research or abandon their work entirely. In contrast, institutional researchers benefit from government fellowships, stipends, and grants that cover their research expenses.

To support independent scholars, India must introduce dedicated fellowships and funding programs that allow them to work without financial barriers.

Proposed Solutions: Introducing Fellowships for Independent Researchers

1. Establishing an "Independent Research Fellowship Program"

- India should launch an Independent Research Fellowship (IRF), modeled after successful global programs like:
- The Humboldt Fellowship (Germany) – funds independent

researchers in science and engineering.

- The Marie Curie Fellowship (European Union) – provides direct research funding to individual scientists.
- The Sloan Research Fellowship (USA) – supports young independent researchers.

The IRF should provide financial support for independent researchers in areas like:

- Basic sciences and engineering.
- Social sciences and humanities.
- AI, quantum computing, and emerging technologies.

2. Monthly Stipends for Independent Researchers

- A ₹50,000–₹1,00,000 monthly stipend should be provided for independent researchers who qualify for national fellowships.
- This stipend would: Cover living expenses, allowing researchers to focus on their projects full-time. Encourage early-career researchers to pursue independent research instead of institutional dependency. Help researchers in fields with limited funding opportunities (e.g., humanities, interdisciplinary studies).

3. Funding for Independent Research Travel and Conference Participation

- Many independent researchers lack funding to attend international conferences, limiting their ability to share findings.
- The government should introduce conference travel grants for independent scholars, ensuring that they can: Present their

work at international forums. Network with global researchers. Gain visibility in academic and industry circles.

- By introducing national-level fellowships and stipends, India can create a thriving ecosystem for independent research.

7.2.3 Developing Mentorship Networks Connecting Senior Academics with Independent Scholars

One of the most significant disadvantages independent researchers face is the lack of academic mentorship. Institutional researchers benefit from faculty guidance, research collaborations, and structured career development, whereas independent scholars often work without expert feedback or mentorship.

To bridge this gap, India must develop structured mentorship programs that connect independent researchers with senior academics, industry experts, and research advisors.

Proposed Solutions: Creating Mentorship Networks for Independent Researchers

1. Establishing a National Mentorship Program for Independent Researchers

A government-backed National Mentorship Program should be introduced to:

- Connect experienced scientists and faculty members with independent researchers.
- Provide one-on-one mentorship and research guidance.
- Offer structured training in research methodology, publishing, and funding applications.

2. Creating an Online Mentorship Platform for Independent Researchers

A digital mentorship portal should be launched where:

- Independent researchers can connect with academic mentors across institutions.
- Senior faculty members can offer guidance in proposal writing, research methodologies, and peer review processes.
- Industry professionals can collaborate with independent researchers on applied research projects.

3. Encouraging Senior Academics to Mentor Independent Researchers

Universities should incentivize professors to mentor independent researchers by:

- Recognizing mentorship contributions in faculty evaluations.
- Providing research collaboration grants for professors who guide independent scholars.
- Encouraging cross-institutional and non-institutional mentorship collaborations.

By developing mentorship networks, independent researchers can receive the guidance, academic support, and career-building opportunities needed to succeed in their fields.

Conclusion

To create a research-friendly environment for independent scholars, India must:

- Open university and government research labs to unaffiliated researchers.
- Introduce fellowships and stipends to financially support independent research.
- Develop structured mentorship networks to connect independent researchers with experts.

These reforms will eliminate the barriers that currently prevent independent researchers from contributing to scientific progress, ensuring that India maximizes its research potential beyond institutional boundaries.

7.3 Encouraging Corporate R&D Funding for Individuals

One of the biggest weaknesses in India's research ecosystem is the lack of corporate funding for independent researchers. While private companies invest heavily in institutional R&D, they rarely fund individual researchers, inventors, or non-affiliated scientists. This is in contrast to global innovation hubs like the United States, Germany, and China, where corporate R&D plays a major role in supporting independent innovators.

If India wants to strengthen its position as a research and innovation leader, it must create a framework that encourages corporate investment in independent research. This can be achieved through:

- Tax incentives for companies that invest in independent researchers.
- Industry-academia collaboration models that include independent scholars.
- Startup incubation programs to support non-institutional innovators.

This section explores how India can develop a strong corporate-backed research funding system that benefits both independent researchers and the private sector.

7.3.1 Tax Incentives for Companies Investing in Independent Research

Most private-sector research funding in India is directed toward institutional R&D labs and universities, with very little support for individual scientists or independent researchers. In contrast, countries like the US, Germany, and South Korea have tax incentive programs that encourage companies to fund external research, including independent innovators.

Challenges in Corporate R&D Funding for Independent Researchers

- Companies prefer investing in institutional projects because universities offer structured partnerships and government co-funding.
- There is no tax benefit for funding individual researchers, making it financially unattractive for companies.
- Independent researchers lack direct access to corporate research grants, which are often tied to institutional collaborations.

Proposed Solutions: Encouraging Corporate Funding Through Tax Benefits

1. Implementing Tax Deductions for Corporate R&D Investments in Independent Researchers

The Indian government should introduce tax deduction policies that allow companies to:

- Claim 100–150% tax deductions for investments in independent research projects.
- Receive additional tax credits for funding early-stage research by non-affiliated scientists.
- Write off expenses related to funding independent R&D projects, lab access, and intellectual property development.

Countries like the US (R&D Tax Credit Program) and the UK (Research and Development Expenditure Credit) have successfully used tax breaks to encourage corporate R&D partnerships with independent researchers.

2. Creating Government-Backed Research Sponsorship Programs

The government should introduce a Public-Private R&D Sponsorship Fund, where:

- Companies contribute a portion of their CSR (Corporate Social Responsibility) funds toward independent research.
- The government matches private investments to encourage long-term corporate engagement in non-institutional R&D.

Independent researchers can apply for funding through a transparent evaluation process.

3. Offering Patent-Linked Tax Incentives for Corporate R&D Support

Companies that fund independent research leading to patents should receive:

- Reduced corporate tax rates on profits earned from innovations funded by independent researchers.
- Exclusive licensing options for patents developed through their funding, ensuring mutual benefit.
- Government recognition and incentives for participating in national innovation programs.

By introducing tax benefits and corporate-sponsored research funds, India can encourage businesses to actively invest in independent researchers, driving innovation outside of traditional institutional settings.

7.3.2 Industry-Academia Collaboration Models That Include Independent Researchers

A major reason corporate R&D investments do not reach independent researchers is the lack of structured industry-academia collaboration models that recognize non-institutional scientists. Most corporate-academic partnerships only involve universities and research institutions, leaving independent researchers excluded from industry-backed innovation projects.

Challenges in Industry-Academia Collaboration for Independent Researchers

- Companies prefer structured institutional partnerships due to existing infrastructure and regulatory clarity.
- Independent researchers are not recognized in corporate-academic R&D programs, preventing them from accessing corporate grants.
- There are no formal platforms connecting independent researchers with industry needs, limiting collaboration opportunities.

Proposed Solutions: Expanding Industry-Academia R&D to Include Independent Researchers

1. Establishing a National "Independent Research-Industry Partnership Program"

India should create a national program connecting independent researchers with industry R&D needs.

This program should:

- Allow independent researchers to apply for corporate-funded research projects in various fields.
- Provide direct industry mentorship and collaboration opportunities for non-affiliated scientists.

- Offer funding incentives to companies that integrate independent researchers into their R&D programs.

2. Creating Open Innovation Platforms for Corporate-Independent Researcher Collaboration

A national online platform should be developed where:

- Companies post R&D challenges and seek solutions from independent researchers.
- Independent researchers submit proposals for funding and implementation.
- Industry and independent scientists co-develop research projects, ensuring knowledge-sharing and practical application.

This model is successfully used by MIT's Industrial Liaison Program (USA) and Fraunhofer Institutes (Germany), enabling direct collaboration between companies and non-institutional researchers.

3. Allowing Independent Researchers to Apply for Industry-Academia Grants

Many industry-academic funding schemes in India (such as SERB-Industry Collaborative R&D Grants) currently exclude independent researchers.

Policy reforms should:

- Expand eligibility criteria to allow independent researchers to apply.
- Encourage joint applications, where independent scholars can collaborate with corporate R&D teams.
- Provide government co-funding for independent researchers working on industry-relevant innovations.

By expanding corporate-academic collaboration models, India can create more opportunities for independent researchers to access industry R&D funding and contribute to private-sector innovation.

7.3.3 Startup Incubation Support for Independent Innovators and Entrepreneurs

Many independent researchers in India work on groundbreaking innovations that could lead to successful startups and technological advancements. However, due to lack of funding, lab access, and business mentorship, many of these ideas fail to reach commercialization.

If India wants to turn independent researchers into successful innovators, it must provide startup incubation support to help them develop their ideas into marketable technologies.

Challenges Faced by Independent Innovators in Startup Development

- Most startup incubators are linked to universities, making it hard for independent researchers to access support.
- There is limited funding for non-institutional deep-tech startups, forcing many researchers to seek foreign investment or abandon their work.
- Independent researchers lack business mentorship, making it difficult to scale their innovations.

Proposed Solutions: Supporting Startups Founded by Independent Researchers

1. Creating Dedicated Startup Incubation Programs for Independent Researchers

India should establish special startup incubators for non-institutional innovators, modeled after:

- Germany's Fraunhofer Institutes, which help independent

researchers commercialize their work.

- The US SBIR (Small Business Innovation Research) Program, which funds deep-tech startups from individual researchers.

These incubators should provide:

- Funding and workspace for product development.
- Business mentorship from industry leaders.
- Legal and patent support for commercializing innovations.

2. Launching a Government-Backed "Independent Innovators Fund"

A national-level Independent Innovators Fund should be created to:

- Provide seed grants of ₹20–50 lakh for independent researcher-led startups.
- Offer low-interest R&D loans for scaling innovation projects.
- Facilitate collaborations between independent innovators and venture capital firms.

3. Expanding Startup India and Make in India Programs to Support Independent Researchers

Current government programs like Startup India and Make in India should:

- Include independent researchers in eligibility criteria.
- Offer fast-track funding for researcher-led startups.
- Connect independent innovators with corporate investors.

By creating startup incubation support, India can turn its independent researchers into successful entrepreneurs, driving self-reliant innovation and economic growth.

Conclusion

For India to fully leverage the potential of independent researchers, it must:

- Introduce tax incentives for companies investing in non-institutional R&D.
- Expand industry-academia collaboration to include independent scholars.
- Provide startup incubation support for researcher-led innovations.

These reforms will bridge the gap between independent research and corporate funding, ensuring sustainable, industry-backed innovation in India.

7.4 Global Best Practices That India Can Adopt

Independent researchers across the world face challenges, but several countries have successfully developed funding models, research ecosystems, and open-access platforms that provide direct support to non-institutional scholars. Countries like China, Germany, and the United States have established structured frameworks that fund, mentor, and integrate independent researchers into mainstream scientific progress.

By learning from global best practices, India can develop a robust, inclusive research ecosystem that empowers independent researchers and maximizes the country's intellectual potential.

7.4.1 How China, Germany, and the US Support Independent Researchers

1. The United States: Government Grants and Private Research Funding

The US has one of the most developed independent research funding systems in the world, providing a mix of government grants, private endowments, and industry-backed initiatives.

Key Initiatives Supporting Independent Researchers in the US

The National Science Foundation (NSF) Grants

- Offers direct research funding to independent scholars across various disciplines.
- Includes special programs like the NSF Small Grants Program, which provides seed funding for early-stage researchers.

The Howard Hughes Medical Institute (HHMI) Investigator Program

- Funds biomedical researchers without requiring institutional affiliation.
- Offers long-term, flexible grants to scientists working on innovative medical research.

The Small Business Innovation Research (SBIR) and Small Business Technology Transfer (STTR) Programs

- Provides funding for independent inventors, scientists, and small-scale researchers to commercialize their work.
- Encourages deep-tech entrepreneurship and industry collaboration.

Lessons for India

- Introduce flexible government research grants that allow independent researchers to apply directly.
- Create fellowships for non-affiliated scientists similar to the

HHMI Investigator Program.

- Encourage private-sector funding for deep-tech innovations through targeted government support.

2. Germany: Research Fellowships and Institutional Independence

Germany has built an independent research-friendly ecosystem by offering well-structured fellowships, open-access funding, and institutional autonomy.

Key Initiatives Supporting Independent Researchers in Germany

The Max Planck Society Fellowship

- Provides funding for researchers outside of universities, allowing them to work independently in specialized research centers.
- Offers international collaborations and full funding support for innovative projects.

The Alexander von Humboldt Foundation Fellowships

- Supports high-achieving independent researchers by providing research grants and lab access at German institutions.
- Includes flexible funding programs that do not require institutional affiliation.

The Fraunhofer Institutes' Industry Collaboration Program

- Facilitates partnerships between independent researchers and industries.
- Provides lab space, corporate funding, and direct research support to non-affiliated scientists.

Lessons for India

- Establish research fellowships for independent scholars, similar to the Max Planck Society and Humboldt Foundation programs.
- Encourage independent research centers that operate outside traditional universities.
- Create government-backed R&D hubs where independent researchers can collaborate with industries.

3. China: Direct Government Support for Independent Researchers

China has aggressively expanded its independent research ecosystem through government-backed funding, industrial research partnerships, and open-access initiatives.

Key Initiatives Supporting Independent Researchers in China

The Thousand Talents Plan

- Attracts and funds top independent researchers globally, offering them full research funding and high salaries.
- Encourages Chinese researchers to return from abroad and conduct independent research.

The National Natural Science Foundation of China (NSFC)

- Offers direct funding to independent scientists working on emerging technologies.
- Includes special grants for early-career researchers and individuals working outside traditional institutions.

The China Academy of Sciences (CAS) Open Research Program

- Provides independent researchers access to labs, libraries, and

data resources.
- Encourages private-sector partnerships for independent scholars.

Lessons for India

- Develop a national talent program that directly funds high-potential independent researchers.
- Provide financial incentives for Indian researchers working abroad to return and conduct independent research.
- Create government-funded open research platforms that allow independent scholars to collaborate with state-backed institutions.

7.4.2 Lessons from Global Open-Access Funding Models

A major global shift in research funding has been the adoption of open-access funding models, ensuring that independent researchers can publish and share their findings freely.

Successful Open-Access Funding Models India Can Learn From

1. The European Research Council (ERC) Grants

- Provides fully-funded open-access research grants to individual scholars.
- Allows researchers to own their intellectual property without institutional restrictions.

2. The Wellcome Trust Open Science Program (UK)

- Offers grant funding for independent researchers to conduct open-access studies.
- Promotes transparent and accessible research publications.

3. The OpenAI Research Collaboration Model (US)

- Funds independent AI researchers through a decentralized grant model.
- Allows public access to datasets, tools, and collaborative research hubs.

Lessons for India

- Develop national open-access research grants, ensuring independent scholars can publish freely.
- Encourage private-sector open-science funding, similar to the Wellcome Trust and OpenAI models.
- Make government-funded research freely accessible, promoting innovation across industries.

7.4.3 Developing an Independent Research Ecosystem Based on International Frameworks

Key Steps India Should Take to Develop an Independent Research Ecosystem

1. Establishing a National Independent Research Fund (NIRF)

- A dedicated fund for independent researchers to apply for government-backed grants.
- Structured funding similar to Germany's Humboldt Fellowship or China's NSFC.

2. Creating Open Labs for Independent Researchers

- Allow unaffiliated researchers to use government-funded laboratories.
- Develop open research centers similar to China's Academy of Sciences model.

3. Introducing a National Independent Researcher Fellowship

- Provide financial stipends for independent researchers to work full-time.
- Modeled after the Max Planck and European Research Council fellowship programs.

4. Encouraging Corporate and Industry Partnerships for Independent Research

- Introduce tax incentives for companies funding independent research.
- Encourage corporate-academic R&D collaborations that include non-institutional scientists.

5. Implementing an Open-Access Publishing Policy

- Mandate that government-funded research be freely available to ensure accessibility.
- Provide funding for independent researchers to publish in high-impact journals.

Conclusion

Countries like the US, Germany, and China have successfully developed funding models, infrastructure, and open-access research ecosystems that support independent researchers.

For India to catch up and compete in global innovation, it must:

- Introduce structured government grants for non-affiliated researchers.
- Develop fellowships and open research labs for independent scholars.
- Encourage corporate R&D investment in independent

research.
- Adopt global open-access funding models to make research freely available.

By learning from international best practices, India can transform its research ecosystem, ensuring that innovation is driven by talent and ideas, not just institutional privilege.

7.5 Challenges in Policy Implementation

While reforming India's research ecosystem to support independent researchers is crucial, implementing these changes comes with significant challenges. The current system is deeply entrenched in institutional privilege, bureaucratic inefficiencies, and a lack of transparency, making it difficult to introduce new funding models, open-access research programs, and corporate-backed R&D initiatives.

For policy reforms to succeed, India must overcome institutional resistance, streamline administrative processes, and ensure accountability in research funding. This section explores the three key barriers to policy implementation and provides recommendations on how to address them effectively.

7.5.1 The Resistance from Institutional Research Monopolies

One of the biggest obstacles to independent research reform is the institutional monopoly over research funding, lab access, and publishing networks. Leading research institutions—such as IITs, IISc, CSIR labs, and central universities—have long benefited from exclusive government grants, infrastructure, and industry partnerships, leaving little room for independent scholars.

How Institutional Research Monopolies Resist Change

1. Exclusive Control Over Government Research Grants

- Government research funding is largely funneled into elite institutions, with independent researchers receiving little to no direct support.

- Institutional researchers often lobby against funding for independent scholars, fearing competition for limited research grants.
- Any attempt to diversify funding access is met with pushback from established academic groups that wish to maintain their privileged status.

2. Restricted Lab and Facility Access for Independent Scholars

- University and government labs are closed to unaffiliated researchers, forcing independent scholars to rely on personal funding for research facilities.
- Institutions resist opening lab access to outsiders, citing concerns over intellectual property, security, and administrative complications.
- This creates a closed ecosystem where research progress is controlled solely by institutions, shutting out non-affiliated innovators.

3. Academic Publishing Networks Favor Institutional Researchers

- High-impact journals often prioritize submissions from institutional researchers, making it difficult for independent scholars to gain recognition.
- Institutional researchers have exclusive access to journal memberships, editorial boards, and peer review panels, creating biases against unaffiliated scholars.

Solutions to Overcome Institutional Resistance

- Mandate a fixed percentage of government research grants to be allocated to independent researchers.

- Legally enforce lab access policies that allow unaffiliated researchers to use government-funded research facilities.
- Introduce open publishing platforms where independent researchers can submit their work without institutional barriers.

By breaking institutional monopolies, India can create a more inclusive research ecosystem that values scientific innovation over institutional privilege.

7.5.2 Administrative and Bureaucratic Inefficiencies in Research Governance

Even when new policies are introduced, their implementation is often delayed due to bureaucratic inefficiencies, complex administrative procedures, and lack of coordination between research agencies. Many funding programs fail to reach independent researchers due to slow processing, excessive paperwork, and rigid application criteria.

Challenges in Bureaucratic Implementation of Research Reforms

1. Lengthy and Complex Research Grant Application Processes

- Many research funding applications require multiple approvals from different government agencies, leading to excessive delays.
- Independent researchers must navigate complex documentation requirements, often without institutional support.
- The absence of a single-window clearance system forces researchers to submit paperwork to multiple offices, slowing down fund disbursement.

2. Inefficient Grant Disbursement and Utilization Policies

- Even when grants are approved, funds are released in multiple stages, causing disruptions in research continuity.
- Bureaucratic delays in budget approvals, expense justifications, and financial audits lead to underutilization of allocated research funds.
- Unlike private-sector funding, government research grants have rigid spending rules, restricting flexibility in how independent researchers manage their projects.

3. Lack of Coordination Between Research Funding Agencies

- India has multiple research funding bodies (SERB, DST, DBT, CSIR, ICAR, DRDO), but their policies are often uncoordinated.
- There is no centralized research funding portal, making it difficult for independent scholars to identify and apply for grants.
- Many funding programs overlap in purpose but lack synergy, leading to inefficiencies in grant allocation.

Solutions to Improve Research Administration Efficiency

- Establish a National Research Funding Portal, where independent researchers can apply for grants through a single-window system.
- Introduce an automatic funding disbursement system to ensure timely release of grants.
- Simplify grant application processes by reducing bureaucratic paperwork and unnecessary approvals.
- Encourage private-public partnerships in research administration, allowing industry stakeholders to streamline funding mechanisms.

By reducing bureaucratic inefficiencies, India can ensure that research funds are accessible, transparent, and efficiently utilized.

7.5.3 Ensuring Transparency and Accountability in New Funding Programs

Corruption, favoritism, and lack of transparency have historically plagued research funding in India, with grants often awarded based on institutional influence rather than research merit. If India introduces new funding models for independent researchers, it must ensure strong oversight mechanisms to prevent misuse, favoritism, and misallocation of funds.

Challenges in Maintaining Transparency and Accountability

1. Favoritism and Nepotism in Research Grant Allocations

- Many research grants are awarded based on academic networks and political connections, rather than on the quality of proposals.
- Institutional researchers have greater lobbying power, often securing funding at the expense of equally deserving independent scholars.

2. Lack of Independent Review Mechanisms for Research Funding

- Current funding review boards are often dominated by institutional researchers, leading to biased decision-making.
- There is no structured, independent evaluation panel to assess grant applications objectively.

3. Misallocation and Underutilization of Research Funds

- Some government-funded research projects remain unfinished or fail to deliver results, leading to wasted resources.

- The absence of proper monitoring systems allows funds to be diverted or underutilized without accountability.

Solutions to Strengthen Transparency in Research Funding

Create an Independent Research Funding Review Committee

- The committee should include experts from diverse fields, ensuring fair and unbiased evaluations.
- It should be free from institutional influence to prevent favoritism in grant approvals.

Implement a Publicly Transparent Research Grant Database

- The database should list all approved research projects, along with their funding amounts, progress updates, and outcomes.
- This will allow public scrutiny and reduce the risk of corruption in research funding.

Introduce Performance-Based Grant Renewals

- Instead of automatic multi-year funding, grants should be reviewed based on research progress.
- Researchers should be required to demonstrate tangible outcomes before receiving additional funding.

By ensuring transparency and accountability, India can prevent misuse of research funds and build a trustworthy funding ecosystem for independent scholars.

Conclusion

While reforming research policies to support independent scholars is essential, successful implementation requires overcoming key barriers.

For policy reforms to be effective, India must:

- Break the institutional monopoly on research funding and infrastructure.
- Reduce bureaucratic inefficiencies in research administration.
- Ensure transparency and accountability in research funding programs.

By addressing these policy implementation challenges, India can create a fair, transparent, and inclusive research ecosystem, ensuring that independent researchers receive the recognition, funding, and support they deserve.

7.6 Legislative Recommendations for India's Research Policy

India's current research policies overwhelmingly favor institutional researchers, leaving independent scholars without structured funding, lab access, or recognition. While policy reforms are necessary to address these imbalances, legislative action is required to institutionalize these changes, ensuring that independent researchers are legally recognized, funded, and integrated into India's research ecosystem.

To make India's research policies inclusive and globally competitive, key legislative measures must be introduced, including:

- Amendments to existing research grant policies to formally include independent scholars.
- The establishment of a National Independent Research Council (NIRC) to oversee non-institutional research funding and policies.
- Legal frameworks for corporate and government collaboration, ensuring sustainable funding mechanisms for independent research.

This section explores these legislative recommendations in detail and how they can help India create a research ecosystem that values innovation over institutional privilege.

7.6.1 Amendments to Existing Research Grant Policies to Include Independent Scholars

Currently, most government-funded research grants in India are restricted to institutional researchers, requiring applicants to be affiliated with universities, IITs, IISc, or government research bodies. This excludes independent researchers from accessing critical funding opportunities.

Key Issues with Current Grant Policies

- Institutional Affiliation Requirement: Independent researchers cannot apply for government grants without an institutional endorsement.
- Limited Eligibility for Non-Academic Scientists: Many grants exclude industry professionals, self-funded scholars, and non-institutional innovators.
- Rigid Grant Disbursement Models: Current policies do not allow for flexible funding structures that accommodate independent research timelines.

Proposed Legislative Amendments

1. Removing Institutional Affiliation as a Mandatory Grant Requirement

- Amend funding guidelines under SERB (Science and Engineering Research Board), DBT (Department of Biotechnology), and DST (Department of Science & Technology) to allow independent researchers to apply for grants without institutional backing.
- Introduce direct-to-researcher funding models, similar to the

US National Science Foundation (NSF) grants that support independent scholars.

2. Expanding Eligibility Criteria for Research Grants

- Modify eligibility rules to include non-affiliated scientists, self-funded researchers, and industry professionals working on independent projects.
- Establish fellowship categories specifically for independent researchers, modeled after Germany's Humboldt Fellowship and the European Research Council (ERC) grants.

3. Introducing Flexible Grant Disbursement Policies

- Implement multi-phase research funding, where independent scholars receive grants based on milestones rather than rigid timelines.
- Allow researchers to retain grant funding across multiple projects, ensuring long-term sustainability.

By amending India's existing research grant policies, the government can immediately expand funding access to thousands of non-affiliated scholars and innovators.

7.6.2 Establishing a National Independent Research Council (NIRC) for Non-Institutional Research

A dedicated agency is needed to oversee funding, mentorship, and infrastructure access for independent researchers in India. This can be done by establishing a National Independent Research Council (NIRC), which will act as a regulatory and funding body for non-institutional scholars.

Why India Needs a National Independent Research Council (NIRC)

- Lack of a dedicated platform for independent researchers to seek funding, mentorship, and lab access.
- Existing research bodies (DST, DBT, SERB, CSIR) focus primarily on institutional grants, leaving non-affiliated scholars unrecognized.
- No structured representation for independent researchers in India's science and technology policies.

Proposed Structure and Functions of the NIRC

1. Funding and Grant Distribution

- Manage a dedicated fund for independent researchers, separate from institutional research grants.
- Oversee peer-reviewed grant applications for non-affiliated scholars, ensuring fair distribution of funds.

2. Infrastructure and Lab Access

- Facilitate access to government-funded labs and research centers for independent scholars.
- Develop shared research spaces, where non-affiliated researchers can collaborate with institutions and industries.

3. Mentorship and Skill Development Programs

- Connect independent researchers with senior scientists, industry experts, and global research networks.
- Provide career guidance, project management training, and publishing support to ensure that independent scholars succeed.

4. Policy Advisory and Representation

- Ensure that independent researchers are represented in national science policy discussions.
- Collaborate with industry, academia, and government agencies to integrate non-institutional research into mainstream scientific progress.

Establishing the NIRC will give independent researchers an institutional voice, ensuring sustained policy support and funding for non-affiliated scholars.

7.6.3 Legal Frameworks to Ensure Corporate and Government Collaboration in Funding Individual Researchers

India must introduce legislative measures to create a structured funding ecosystem where corporate and government agencies collaborate to support independent researchers.

Current Barriers in Corporate-Government Research Collaboration

- Lack of tax incentives for companies funding independent research.
- Absence of structured industry-academia funding models that include non-affiliated researchers.
- No legal framework ensuring transparent corporate investments in independent R&D projects.

Proposed Legislative Measures

1. Tax Incentives for Corporate R&D Investment in Independent Researchers

- Amend the Income Tax Act, 1961, to provide 150% tax deductions for companies funding independent research.
- Offer reduced corporate tax rates for businesses that sponsor

non-institutional research projects.

- Provide research investment credits for corporate-backed independent R&D projects, modeled after the US R&D Tax Credit Program.

2. Establishing a Public-Private Research Funding Model

- Introduce a Public-Private Independent Research Fund (PPIRF), where:
- Companies co-invest with the government in research grants for independent scholars.
- Independent researchers can apply for joint government-corporate funding without institutional affiliation.
- Modeled after China's National Science and Technology Support Program, which combines state and corporate funding for independent research.

3. Legal Mandates for Industry-Academia Collaboration that Includes Independent Researchers

- Amend India's Industrial R&D Policy to require companies to allocate a percentage of R&D budgets for independent research.
- Create legal provisions for independent researchers to partner with corporate R&D labs, ensuring access to facilities and funding.
- Encourage international collaboration agreements, allowing Indian independent researchers to access foreign research grants and industry partnerships.

By creating strong legal frameworks, India can institutionalize corporate-government research collaborations, ensuring sustained financial and infrastructure support for independent researchers.

Conclusion

For India to become a global leader in scientific research, legislative changes are essential to integrate independent researchers into the national R&D framework.

Key legislative recommendations include:

- Amending research grant policies to allow direct funding for non-affiliated scholars.
- Establishing a National Independent Research Council (NIRC) to oversee funding, infrastructure, and mentorship for independent researchers.
- Implementing legal frameworks for corporate-government R&D collaboration, ensuring long-term funding sustainability.

By enacting these legislative reforms, India can create a research ecosystem where innovation thrives beyond institutional boundaries, allowing independent researchers to contribute to national and global scientific progress.

7.7 Case Studies: Successful Independent Research Funding Models

While independent researchers face significant barriers in many countries, some global institutions and funding models have successfully supported non-institutional scholars through direct research grants, open-access programs, and industry collaborations. These case studies demonstrate how structured funding ecosystems can empower independent researchers, drive scientific breakthroughs, and establish global research leadership.

By analyzing these models, India can develop a sustainable and inclusive framework that supports independent research and innovation.

7.7.1 The Howard Hughes Medical Institute (HHMI), USA: Direct Funding for Independent Biomedical Researchers

Overview

The Howard Hughes Medical Institute (HHMI) is one of the most prestigious private biomedical research funding organizations in the world. Unlike traditional grant programs, HHMI provides long-term, flexible funding to individual scientists, allowing them to conduct high-risk, high-reward research without institutional restrictions.

Key Features of HHMI's Funding Model

1. Direct Support for Individual Researchers

- HHMI's Investigator Program funds independent researchers based on scientific potential, not institutional affiliation.
- Researchers receive five- to seven-year funding, ensuring stability and research freedom.

2. No Institutional Control Over Research

- Unlike university-based grants, HHMI does not require researchers to follow institutional policies.
- Scientists retain intellectual property rights over their discoveries.

3. Focus on Innovative and Risk-Taking Research

- HHMI prioritizes groundbreaking research that institutions often overlook due to funding risks.
- This approach has led to major discoveries in genetics, cancer research, and neuroscience.

Lessons for India

- Establish a National Independent Research Fellowship similar to HHMI's Investigator Program.
- Provide long-term research grants that allow independent scientists to work without rigid institutional oversight.
- Prioritize innovation-driven research funding, focusing on high-impact, unconventional projects.

7.7.2 The European Research Council (ERC) Grants: Supporting Individual Excellence in Frontier Research

Overview

The European Research Council (ERC) provides one of the most successful independent research funding models globally, supporting individual scientists across disciplines through competitive, merit-based grants.

Key Features of the ERC Model

1. Grants for Individual Excellence, Not Institutional Affiliation

- ERC funds independent researchers at different career stages, from early-career scientists to established researchers.
- Grants are awarded based on research quality, without requiring university or institutional affiliation.

2. Open-Access Funding Across All Scientific Fields

- ERC grants support research in science, engineering, humanities, and social sciences.
- Researchers have full freedom to choose their research topics, ensuring academic independence.

3. International Research Mobility and Collaboration

- ERC encourages researchers to work across countries,

promoting global collaboration and innovation.

- Scientists can carry their funding across different institutions without losing grant support.

Lessons for India

- Establish an Independent Research Grant Program modeled after ERC.
- Encourage cross-border research collaborations by allowing Indian researchers to apply for global funding.
- Introduce open research funding models that prioritize independent scientific inquiry.

7.7.3 China's Thousand Talents Plan: Attracting and Funding Top Independent Researchers Globally

Overview

China's Thousand Talents Plan is a strategic initiative aimed at attracting global scientific talent and funding independent researchers, making China a global leader in innovation.

Key Features of China's Thousand Talents Plan

1. Direct Government Funding for Independent Researchers

- Provides high salaries, lab funding, and research grants to scientists worldwide.
- Supports independent researchers without requiring permanent university affiliation.

2. State-Backed Industry Collaborations

- Connects researchers with government, corporate R&D labs, and technology startups.
- Provides access to high-end research facilities, ensuring infrastructure support.

3. Encouraging Talent Return and Retention

- Offers competitive funding to Chinese researchers working abroad, bringing them back to China.
- Provides long-term fellowships for foreign researchers interested in working in China.

Lessons for India

- Create a national talent attraction program that funds Indian researchers working abroad.
- Encourage industry-academia collaboration by integrating independent researchers into corporate R&D projects.
- Offer competitive financial packages to retain top Indian scientists in India.

7.7.4 The Max Planck Institute, Germany: Open-Access Research and Independent Fellowships

Overview

The Max Planck Society is Germany's leading research institution, providing fully funded fellowships to independent researchers and promoting open-access scientific research.

Key Features of Max Planck's Model

1. Independent Research Institutes Without University Control

- Max Planck institutes operate autonomously, allowing scientists to work without university bureaucracy.
- Researchers receive full funding and lab access, ensuring research independence.

2. Open-Access Funding for Non-Affiliated Scientists

- Fellowships are available to both German and international researchers, regardless of institutional ties.
- Funding is provided for basic and applied research across disciplines.

3. Industry and Government Research Collaborations

- Encourages joint research between independent scientists, universities, and corporate R&D.
- Provides open lab access to researchers, ensuring infrastructure support.

Lessons for India

- Establish government-funded independent research institutes, allowing non-affiliated scholars to work freely.
- Promote open-access research fellowships, similar to the Max Planck model.
- Encourage joint research collaborations between independent scientists, industries, and universities.

7.7.5 India's ISRO Model: How Independent Thinking Led to a Globally Recognized Space Program

Overview

The Indian Space Research Organisation (ISRO) is an example of how independent thinking, strategic government support, and minimal bureaucracy can lead to world-class research achievements.

Key Features of ISRO's Research Success

1. Research Freedom and Minimal Bureaucratic Interference

- Unlike many other government research agencies, ISRO operates with minimal bureaucratic control, allowing scientists to take risks and innovate.

- Research programs are goal-driven and efficiently managed, ensuring rapid progress.

2. Direct Government Funding Without Institutional Barriers

- ISRO receives direct research funding from the Indian government, without having to navigate academic grant systems.
- Scientists are given long-term project-based funding, ensuring stability.

3. Open Collaboration Between Industry and Research Teams

- ISRO partners with private companies to develop new technologies, creating a sustainable research ecosystem.
- Encourages cross-institutional and independent collaborations, integrating multiple scientific fields into space research.

Lessons for India

- Reduce bureaucratic delays in research funding, following ISRO's efficient model.
- Encourage direct research funding programs for independent scientists.
- Promote mission-driven research funding, ensuring accountability and impact.

Conclusion

These case studies highlight successful global approaches to independent research funding and provide valuable lessons for India.

Key Takeaways for India

- Develop flexible, long-term funding programs for independent researchers (HHMI, ERC, Max Planck).
- Create national talent programs to attract and retain top researchers (Thousand Talents Plan).
- Reduce bureaucratic barriers and provide direct research funding (ISRO model).
- Encourage corporate-academic collaboration to integrate independent researchers into the industry.

By adopting these global best practices, India can establish a thriving independent research ecosystem, ensuring that scientific innovation is not limited to institutional boundaries.

8. Crowdsourcing and Private Funding: A Possible Alternative?

In the previous chapters, we explored the systemic barriers that prevent independent researchers in India from accessing government grants, institutional resources, and structured funding opportunities. Despite the urgent need for reforms, policy changes often take years to materialize, leaving independent scholars to either self-fund their research or abandon it entirely. This has led to an increasing reliance on alternative funding models, particularly crowdsourcing, private research grants, and startup-driven research initiatives.

Globally, crowdfunding platforms, private foundations, corporate R&D sponsorships, and venture capital-backed innovation hubs have played a significant role in financing independent scientific research. These models provide greater flexibility, quicker access to funds, and the ability to work outside bureaucratic constraints. However, they also raise ethical concerns, particularly regarding corporate influence on research outcomes, the commercialization of scientific discoveries, and the risk of prioritizing marketable innovations over fundamental research.

This chapter explores the potential of private and crowdsourced funding as an alternative to government-backed research grants, examining successful case studies, opportunities, challenges, and the role of policy frameworks in ensuring ethical and independent scientific progress.

8.1 The Rise of Crowdfunding for Scientific Research

Traditional research funding has long been dominated by government grants, institutional budgets, and corporate sponsorships. However, for independent researchers—who often lack institutional affiliation or access to structured funding—crowdfunding has emerged as a revolutionary alternative.

Crowdfunding allows scientists to bypass bureaucratic hurdles and seek direct financial support from the public. Through platforms like Kickstarter, Experiment.com, and GoFundMe, researchers can pitch their ideas to a global audience, attracting small contributions from thousands of individuals who believe in their work. This decentralized funding model has already enabled significant scientific breakthroughs across fields such as medicine, environmental science, and space research.

However, while crowdfunding offers a promising alternative, it is not without its challenges. The need for public appeal, competition with consumer-focused campaigns, and unpredictable funding cycles can make it a risky and inconsistent funding source for researchers.

This section explores how crowdfunding platforms are reshaping research funding, presents case studies of successful crowdfunded scientific projects, and examines the limitations of this model for independent researchers.

8.1.1 How Crowdfunding Platforms Are Reshaping Research Funding

Crowdfunding is fundamentally changing how research is financed by offering an open and democratic funding model. Unlike traditional grants that require lengthy application processes and institutional backing, crowdfunding allows any researcher with a compelling idea to raise funds directly from the public.

Key Ways Crowdfunding is Transforming Research Funding

1. Direct Public Engagement in Science

- Crowdfunding enables researchers to connect with the general public and explain the real-world impact of their work.
- Unlike government-funded projects, which are often confined to academic circles, crowdfunded research encourages direct participation from non-experts, leading to greater public interest in science.

2. Funding Niche and High-Risk Research Projects

- Many groundbreaking but unconventional research ideas struggle to receive institutional funding due to their perceived risk.
- Crowdfunding allows researchers to pursue projects that do not fit within traditional funding categories but still hold scientific value.

3. Faster Fundraising Without Bureaucratic Delays

- Applying for government grants can take months or even years, with uncertain outcomes.
- Crowdfunding eliminates bureaucratic delays, allowing researchers to start their work as soon as they meet their funding goals.

4. Increased Visibility for Independent Researchers

- Traditional funding models often favor institutional

affiliations, making it difficult for non-affiliated researchers to gain recognition.

- Crowdfunding platforms help independent researchers gain credibility through public backing and media attention.

Popular Crowdfunding Platforms for Research

- **Experiment.com** – A science-specific crowdfunding platform dedicated to research projects.
- **Kickstarter & Indiegogo** – Primarily used for creative and tech projects but increasingly funding science-based innovations.
- **GoFundMe** – Frequently used for medical and environmental research projects.
- **Patreon** – Provides ongoing financial support for researchers and science communicators through a subscription-based model.

While these platforms democratize access to research funding, their success depends on public interest, marketing strategies, and the ability to effectively communicate scientific ideas.

8.1.2 Successful Case Studies: Scientific Breakthroughs Funded by the Public

Several groundbreaking research projects have successfully leveraged crowdfunding, demonstrating the potential of public-driven scientific investment.

1. The OSIRIS-REx Space Mission – NASA & Citizen Science Collaboration

- NASA partnered with the public to fund portions of the OSIRIS-REx mission, which successfully collected asteroid samples in 2023.

- The campaign engaged citizen scientists, allowing donors to participate in data analysis and mission updates.

2. Glow-in-the-Dark Plants – A Synthetic Biology Crowdfunding Success

- A team of bioengineers raised over $484,000 on Kickstarter to develop genetically modified plants that glow in the dark.
- This project demonstrated how the public is willing to fund futuristic scientific experiments outside institutional boundaries.

3. The OpenWorm Project – Mapping a Complete Organism's Brain

- Scientists used crowdfunding to digitally simulate the neural network of a simple worm (C. elegans).
- This project helped advance artificial intelligence and neuroscience research, funded entirely by public contributions.

4. The SeaBin Project – Cleaning the World's Oceans

- A small team of environmental engineers developed a floating bin that removes plastic waste from oceans, raising over $350,000 through crowdfunding.
- The project gained global media attention and attracted partnerships with marine conservation groups.

These case studies show that crowdfunding can successfully finance independent research, especially when projects focus on high-impact, public-interest topics.

8.1.3 Challenges and Limitations of Crowdfunding for Independent Researchers

Despite its potential, crowdfunding is not a guaranteed solution for all independent researchers. Several challenges and limitations prevent it from being a reliable primary funding source.

1. Dependence on Public Appeal Over Scientific Merit

- Success on crowdfunding platforms often depends on how well a project is marketed rather than its scientific importance.
- Research that lacks visual appeal or immediate real-world applications struggles to attract public support.

2. Uncertain and Unpredictable Funding

- Unlike structured government or institutional grants, crowdfunding does not guarantee success.
- Many campaigns fail to reach their funding goals, leaving projects underfunded or incomplete.

3. Limited Support for Fundamental Research

- Fundamental scientific research—especially in theoretical physics, mathematics, and abstract disciplines—rarely attracts public donations.
- People are more likely to support tangible, applied research rather than theoretical advancements.

4. High Effort Required for Campaign Success

- Running a successful crowdfunding campaign requires significant time and effort, including:
- Developing high-quality promotional materials (videos, graphics, social media campaigns).

- Engaging with potential backers through Q&A sessions and project updates.
- Managing reward systems and donor incentives, which can distract from actual research.

5. No Long-Term Stability for Researchers

- Crowdfunding provides one-time funding rather than continuous financial support.
- Researchers often struggle to secure recurring funding, making long-term projects difficult to sustain.

6. Potential for Ethical Issues and Misinformation

- Some researchers may overhype their results to attract funding, leading to scientific misinformation.
- Unlike peer-reviewed funding bodies, crowdfunding does not require rigorous research validation, increasing the risk of fraudulent campaigns.

Addressing These Challenges

- To maximize the benefits of crowdfunding while minimizing its risks, researchers can:
- Combine crowdfunding with traditional grants, using public funding to cover gaps in research budgets.
- Target niche communities interested in specific scientific areas (e.g., environmental activists, AI enthusiasts).
- Develop long-term funding strategies, such as using Patreon for sustained monthly donations.
- Partner with educational institutions and influencers to improve campaign visibility.

Conclusion

Crowdfunding offers a powerful alternative to traditional research funding, enabling independent researchers to raise funds directly from the public. While some projects have successfully secured large-scale public funding, not all research fields are suited for crowdfunding, and financial unpredictability remains a major challenge.

For independent researchers, crowdfunding should be seen as a complementary funding source rather than a primary alternative to government or institutional grants. By leveraging crowdfunding alongside private research grants and startup partnerships, scientists can overcome funding limitations and bring innovative ideas to life.

8.2 How Private Research Grants Can Change the Landscape

While government-funded grants have traditionally been the primary source of research financing, private research grants are emerging as a powerful alternative, particularly for independent researchers. Unlike bureaucratically controlled state funding, private foundations, philanthropic organizations, and corporate research initiatives provide more flexible funding opportunities, fewer administrative barriers, and greater autonomy in research direction.

Private research grants have transformed scientific innovation in various fields, from biotechnology and artificial intelligence to climate science and space exploration. They not only support high-risk, high-reward projects but also fill the gaps where public funding falls short. However, reliance on private grants also brings challenges, such as concerns over accountability, ethical oversight, and corporate influence on research outcomes.

This section explores the growing role of private foundations in funding research, the advantages and trade-offs compared to government grants, and how philanthropy and industry can better support independent researchers.

8.2.1 The Growing Role of Private Foundations in Funding Independent Research

Private foundations have become significant players in global research funding, providing grants and fellowships that allow scientists to pursue independent work outside institutional constraints. Unlike government agencies, which often favor incremental and safe research, private funding bodies are more willing to take risks on breakthrough innovations.

Major Private Foundations Supporting Independent Research

1. The Wellcome Trust (UK) – Advancing Biomedical Research

- One of the world's largest research funding organizations, focusing on health, genetics, and life sciences.
- Provides unrestricted grants, allowing researchers to explore bold ideas without institutional oversight.

2. The Howard Hughes Medical Institute (HHMI) (USA) – Supporting Investigator-Led Research

- Funds independent scientists directly, offering long-term grants without rigid institutional ties.
- Emphasizes exploratory and disruptive research that traditional grants often reject.

3. The Gates Foundation – Tackling Global Health Challenges

- Focuses on public health and medical innovation, funding independent researchers working on vaccines, disease eradication, and healthcare accessibility.
- Partners with universities, startups, and NGOs, offering collaborative grants for interdisciplinary research.

4. The Chan Zuckerberg Initiative (USA) – AI and Neuroscience Research

- Supports AI-driven research in biology and medicine, funding non-institutional teams and independent scientists.
- Encourages open science, requiring research to be publicly accessible.

5. Tata Trusts (India) – Empowering Research in Social Sciences and Technology

- Provides grants to Indian researchers focusing on education, health, and sustainable technology.
- Supports independent researchers working outside the formal academic system.

These foundations are redefining research funding, proving that scientific progress does not have to be restricted by government bureaucracy.

8.2.2 Comparison with Government-Funded Grants: Flexibility vs. Accountability

While both government and private funding play crucial roles in research, they differ significantly in terms of structure, flexibility, and accountability.

Key Differences Between Government and Private Research Grants

Factor	Government Grants	Private Research Grants
Application Process	Highly bureaucratic, lengthy, and complex.	Streamlined, with fewer administrative barriers.
Funding Approval	Based on institutional ties and political influence.	Based on scientific merit and innovation potential.
Research Freedom	Limited; projects must align with national priorities.	High; researchers have more autonomy.
Risk Appetite	Favors incremental, low-risk research.	Willing to fund high-risk, breakthrough innovations.
Transparency & Oversight	Requires extensive reporting and government monitoring.	Less regulated; depends on foundation policies.
Industry Collaboration	Rarely supports corporate partnerships.	Often encourages cross-industry collaborations.
Eligibility for Independent Researchers	Limited; preference for university-affiliated scientists.	More inclusive; allows non-institutional researchers to apply.

While private grants offer greater flexibility, they also raise concerns over ethical transparency and lack of public accountability, which are important considerations for researchers and policymakers alike.

8.2.3 Encouraging Philanthropists and Corporate Entities to Support Individual Researchers

Despite the growing role of private funding, independent researchers still face difficulties accessing corporate and philanthropic grants. Many private-sector R&D investments are channeled into institutional partnerships rather than individual-led projects. Encouraging wealthy philanthropists, technology leaders, and private enterprises to directly support independent scientists can bridge the funding gap and accelerate groundbreaking discoveries.

Strategies to Expand Private Funding for Independent Research

1. Creating Tax Incentives for Private Research Funding

- Governments can introduce tax deductions and financial incentives for corporations and philanthropists who invest in independent research.

- Modeled after the US R&D Tax Credit Program, which rewards private companies for funding innovative, non-institutional projects.

2. Establishing Research Endowments and Fellowships

- Private organizations can set up long-term fellowships dedicated to supporting non-affiliated researchers.
- For Example: Google's AI for Social Good Grant, which funds independent AI research for global impact.

3. Developing Open-Access Research Networks

- Private investors and foundations can support decentralized research platforms, allowing independent scientists to collaborate without institutional gatekeeping.
- For Example: The Open Philanthropy Project, which provides grants for open-science initiatives.

4. Encouraging Corporate-Backed Research Incubators

- Large companies can offer lab space, technical support, and mentorship to independent researchers working on projects aligned with their industry.
- For Example: Tesla and SpaceX have funded independent researchers working on battery technology and space propulsion.

5. Promoting Science Philanthropy Among High-Net-Worth Individuals

- Philanthropists like Elon Musk, Jeff Bezos, and Ratan Tata have funded private space, AI, and healthcare research.

- Governments and scientific bodies can actively engage with billionaires and tech entrepreneurs, encouraging them to invest in open scientific research.

By expanding corporate and philanthropic involvement, India can diversify research funding sources, ensuring that independent scientists are not left behind in the knowledge economy.

Conclusion

Private research grants offer a viable alternative to traditional government funding, providing greater autonomy, faster approval processes, and a willingness to support breakthrough innovations. While leading global foundations have successfully funded independent research, India needs to expand its private funding ecosystem by:

- Encouraging philanthropic investment in science and technology.
- Introducing tax incentives for corporate R&D funding.
- Developing open-access, private-funded research fellowships.
- Establishing corporate-backed research incubators for independent scientists.
- Promoting collaboration between industry, private foundations, and non-institutional researchers.

By leveraging private capital and philanthropic generosity, India can unlock new possibilities for independent research, ensuring that scientific progress is not constrained by bureaucracy or institutional privilege.

8.3 The Role of Startups and Innovation Hubs

With traditional research funding models often favoring institutional scientists, independent researchers are increasingly turning to startups, incubators, and venture capital-backed innovation hubs for support. Unlike government grants, which tend to prioritize incremental advancements, deep-tech startups, innovation accelerators, and private investors are more willing to fund high-risk, high-reward scientific research that has the potential to disrupt industries.

Startups are not only funding independent researchers but also hiring them directly to work on cutting-edge projects in fields such as artificial intelligence, biotechnology, nanotechnology, and space exploration. Meanwhile, incubators and accelerators provide resources like lab space, mentorship, and early-stage funding, allowing non-affiliated researchers to turn their ideas into viable businesses.

This section explores how deep-tech startups are driving research funding, the role of incubators and accelerators in supporting non-institutional researchers, and the growing involvement of venture capital in research-driven entrepreneurship.

8.3.1 How Deep-Tech Startups Are Funding Independent Research

Deep-tech startups focus on scientific and engineering innovations rather than consumer-driven applications. These companies often emerge from academic research projects but are increasingly becoming a key funding source for independent researchers.

Why Startups Are Investing in Independent Research

- Faster Innovation Cycles: Startups thrive on rapid technological advancements, making them eager to fund independent research that can give them a competitive edge.
- Flexible Research Models: Unlike universities, startups can fund high-risk, unconventional ideas that traditional grant agencies may reject.

- Immediate Commercialization Potential: Startups focus on real-world applications, which means researchers working with them can quickly translate their ideas into market-ready solutions.

Examples of Startups Funding Research

1. OpenAI (USA) – Funding Independent AI Researchers

- Originally started as a non-profit, OpenAI has funded independent AI scientists working on machine learning, neural networks, and ethical AI.
- Its research has influenced global AI policies and industry standards.

2. SpaceX and Blue Origin – Supporting Aerospace Innovations

- These space startups have hired and funded independent aerospace researchers working on propulsion systems, space travel, and satellite communications.
- Many former NASA scientists and independent physicists have found research opportunities in private space exploration ventures.

3. Biotech Startups in India – Advancing Medical Research

- Indian startups like Bugworks Research, Bharat Biotech, and String Bio are funding independent scientists working on antibiotic resistance, vaccine development, and synthetic biology.
- Many biotech startups provide lab access and research funding for independent molecular biologists and bioengineers.

Challenges for Independent Researchers in Startup Collaborations

- **Risk of Losing Intellectual Property (IP):** Researchers working with startups must carefully negotiate IP ownership to retain rights over their discoveries.
- **Market-Driven Research Focus:** Startups prioritize commercial viability, which may limit funding for pure theoretical research.
- **Short-Term Funding Cycles:** Unlike government grants, which provide long-term support, startup funding often depends on business viability and investor interest.

Despite these challenges, deep-tech startups have opened up new pathways for independent researchers, allowing them to work on cutting-edge technologies without institutional constraints.

8.3.2 Incubators and Accelerators: Providing Resources for Non-Affiliated Researchers

Startup incubators and accelerators play a critical role in bridging the gap between independent research and commercialization. They provide:

- Early-stage funding.
- Access to lab facilities and equipment.
- Business mentorship to help researchers convert their ideas into viable startups.

The Difference Between Incubators and Accelerators

Feature	Incubators	Accelerators
Purpose	Help researchers develop business models and refine technology.	Scale up existing research startups.
Duration	Long-term (1–3 years).	Short-term (3–6 months).
Funding Type	Grants, seed capital, or shared-equity partnerships.	Venture capital investment or equity-based funding.
Focus Area	Deep-tech, biotech, AI, hardware innovations.	Growth-stage startups ready for commercialization.

Examples of Incubators and Accelerators Supporting Research

1. Indian Institute of Science (IISc) Incubation Centre – Nurturing Deep-Tech Startups

- Supports independent researchers working on AI, robotics, and clean energy technologies.
- Provides lab facilities, mentorship, and industry collaborations.

2. SINE (IIT Bombay) – Research-Driven Startup Incubator

- Offers funding and infrastructure support to independent scientists developing new technologies.
- Has incubated over 200 deep-tech startups in India.

3. Y Combinator (USA) – Accelerating Science-Based Startups

- Provides early-stage funding and mentorship to independent AI and biotech researchers.
- Has helped launch companies like Ginkgo Bioworks (synthetic biology) and Atomwise (AI-driven drug discovery).

How Independent Researchers Can Leverage Incubators and Accelerators

- Apply for incubator fellowships that provide access to lab space, funding, and industry mentorship.

- Partner with startup founders who are looking for research-driven innovations.
- Use accelerator programs to pitch research ideas to venture capitalists for scaling up research-based businesses.

Incubators and accelerators bridge the gap between research and commercialization, ensuring that scientific discoveries do not remain trapped in academic silos.

8.3.3 The Potential for Venture Capital and Angel Investors to Support Research-Driven Entrepreneurship

While venture capital (VC) funding is traditionally associated with high-growth startups, there is a growing trend of VCs and angel investors backing research-driven companies.

Why Investors Are Funding Research-Based Startups

- Emerging Technologies Offer High Returns – Investors see deep-tech and biotech startups as the next major investment frontier.
- Long-Term Competitive Advantage – Research-driven companies create disruptive innovations, providing sustained market leadership.
- Government Incentives for Research-Based Startups – Many countries, including India, offer tax benefits for VC-backed scientific startups.

Key Investment Trends in Research-Based Startups

1. AI and Quantum Computing Startups Attracting VC Funding

Companies like DeepMind (AI) and Rigetti Computing (quantum computing) received millions in VC funding to scale their research.

2. Biotech and Pharma Startups Securing Private Investments

mRNA vaccine startups like BioNTech were funded by venture capital before partnering with Pfizer for COVID-19 vaccine production.

3. Clean Energy and Climate Tech Startups Gaining Investor Interest

Indian startups like Ionic Materials (solid-state batteries) and Carbon Clean Solutions (CO2 capture technology) have received funding from angel investors and government-backed VC funds.

How Independent Researchers Can Attract VC and Angel Investment

- Develop research projects with clear commercial applications.
- Pitch research ideas to deep-tech investment firms and venture funds.
- Join startup incubators to refine business models and investor presentations.

While VC funding provides significant financial support, researchers must be aware that investors expect commercial viability, meaning pure academic research may not always attract funding.

Conclusion

Startups, incubators, and venture capital funding have emerged as key alternatives to traditional government research grants. By leveraging startup ecosystems, research incubators, and private investments, independent researchers can:

- Secure funding for high-risk, high-reward projects through deep-tech startups.
- Access lab space, mentorship, and business training via incubators and accelerators.
- Attract venture capital and angel investments to scale

research-driven businesses.

By integrating entrepreneurial approaches into scientific research, independent scholars can gain financial independence and make groundbreaking discoveries outside traditional institutions.

8.4 Ethical Concerns and Risks of Privatized Research Funding

While private funding through corporations, venture capital, and philanthropy offers an alternative to bureaucratic government grants, it also raises serious ethical concerns. When profit-driven entities fund research, there is a risk of corporate influence shaping research priorities, directing outcomes, and even suppressing unfavorable findings.

Additionally, private funding often prioritizes marketable innovations, which means fundamental research in pure sciences, philosophy, and theoretical disciplines may receive little to no support. Without proper regulatory oversight and ethical guidelines, privatized funding could jeopardize the integrity of scientific research, leading to bias, misinformation, and conflicts of interest.

This section examines the ethical challenges of privatized research funding, the risk of commercialization overshadowing fundamental science, and the importance of transparency and independence in privately funded research.

8.4.1 Corporate Influence on Research Priorities and Outcomes

The Power of Industry-Backed Research

When corporations fund research, they often have specific business interests in mind. Unlike government funding—which typically supports a wide range of topics—corporate-backed research tends to focus on areas that offer commercial returns.

Ways Corporate Funding Can Influence Research

- **Selective Funding:** Companies fund projects that align with their business strategies, leaving less profitable but scientifically important research unfunded.
- **Suppression of Unfavorable Results:** If research findings challenge a corporation's product safety, policies, or public image, there is a risk that results may be modified, suppressed, or retracted.
- **Shaping Research Narratives:** Corporations may exaggerate or distort findings to favor their products.
- **Prioritizing Product Development Over Scientific Discovery:** Industry-backed researchers often focus on creating marketable products rather than advancing scientific knowledge.

Examples of Corporate Influence on Research

1. The Tobacco Industry's Influence on Health Research

For decades, tobacco companies funded misleading research to downplay the health risks of smoking. Internal documents later revealed that corporations pressured scientists to manipulate findings and even discredited researchers who exposed the dangers of tobacco.

2. Pharmaceutical Funding and Bias in Drug Research

Many pharmaceutical companies fund clinical trials for their own drugs, which can create bias. Studies have found that industry-sponsored drug trials are more likely to report positive results compared to independent trials.

3. Tech Companies Controlling AI Ethics Research

Big Tech firms like Google and Facebook fund AI ethics research, but some experts argue that these companies use funding to influence discussions in their favor, downplaying concerns about AI surveillance and data privacy.

Mitigating Corporate Influence

- Establish independent review boards to evaluate corporate-funded research.

- Require mandatory disclosure of corporate funding sources in published research papers.
- Ensure corporations cannot prevent researchers from publishing negative results.

While corporate funding can drive technological advancements, without ethical safeguards, it poses a risk to scientific integrity and public trust in research.

8.4.2 The Risk of Research Commercialization Overshadowing Fundamental Science

While applied research leads to technological advancements and market-ready innovations, fundamental research focuses on understanding basic principles—which is the foundation of long-term scientific progress.

Why Fundamental Research is Often Neglected in Private Funding

- **No Immediate Commercial Gains:** Fundamental research does not always lead to immediate profits, making it less attractive for private investors.
- **Long-Term Payoff vs. Short-Term ROI:** Startups and corporations prefer funding research that yields quick commercial applications rather than slow, theoretical advancements.
- **Shift in Research Priorities:** Many research institutions, influenced by private funding, have reduced their focus on fundamental sciences to prioritize market-driven projects.

Case Studies of Commercialization Overriding Fundamental Science

1. The Decline of Pure Mathematics Research

Many tech-driven companies focus on AI, cybersecurity, and quantum computing, leaving disciplines like pure mathematics underfunded—despite its long-term significance in technological advancements.

2. The Space Industry's Shift Toward Commercialization

While NASA historically focused on fundamental space exploration research, the rise of private space companies like SpaceX and Blue Origin has shifted focus toward commercial space tourism and satellite launches, reducing investment in long-term space science research.

3. Pharmaceutical Industry Favoring Profitable Treatments Over Cures

Many pharmaceutical companies prioritize funding chronic disease treatments over one-time cures, as ongoing treatments provide continuous revenue streams, whereas curative solutions do not.

Balancing Commercial and Fundamental Research

- Governments must allocate specific grants for non-commercial, theoretical research.
- Private donors and philanthropic organizations should fund basic sciences without requiring commercial outcomes.
- Universities should create independent research funds that allow scientists to pursue non-market-driven inquiries.

Over-commercialization can stall scientific breakthroughs by focusing only on short-term profitability. A balanced approach is necessary to ensure that fundamental research continues to thrive.

8.4.3 Ensuring Transparency and Independence in Privately Funded Research

Why Transparency Matters in Private Research Funding

Without proper oversight, private research funding can lead to conflicts of interest, research bias, and suppression of unfavorable results. Transparency ensures that scientific integrity is maintained and research findings are trustworthy.

Ethical Risks in Privately Funded Research

- Ghostwriting and Undisclosed Conflicts of Interest: Some companies fund research but pressure scientists to conceal corporate involvement.
- Manipulated Study Designs: Private funders may influence how studies are conducted, leading to biased results.
- Lack of Independent Oversight: Many privately funded research projects are not subjected to rigorous peer review, reducing scientific accountability.

Best Practices for Ensuring Transparency in Private Research Funding

1. Mandatory Disclosure of Funding Sources

- All research publications should clearly state who funded the study and whether there were any conflicts of interest.
- Journals should reject papers that do not disclose financial backers.

2. Establishing Independent Ethics Committees

- Universities and research institutions should have ethics committees that oversee corporate-funded projects.
- These committees should ensure that funding agreements do not interfere with research outcomes.

3. Creating Publicly Accessible Research Databases

- Governments should develop open-access repositories where all corporate-funded research data is made publicly available.

- Transparency in funding sources can help detect and prevent research bias.

4. Strengthening Whistleblower Protections for Scientists

- Scientists should have legal protections if they expose corporate manipulation of research.
- Universities and institutions should have anonymous reporting mechanisms to address unethical practices in privately funded research.
- By implementing these measures, India can ensure that private funding supports scientific progress without compromising ethical integrity.

Conclusion

While private research funding is a valuable alternative to bureaucratic government grants, it comes with significant ethical risks that must be addressed.

Key Takeaways:

- Corporate influence can shape research priorities, potentially leading to bias and suppression of negative results.
- Commercialization often overshadows fundamental research, risking long-term scientific progress.
- Ensuring transparency and independence in research funding is critical to maintaining scientific credibility.

To safeguard scientific integrity, governments, institutions, and private entities must work together to establish:

- Independent ethics review boards for privately funded research.
- Mandatory funding disclosures in research publications.

- A balanced approach that supports both commercial and fundamental research.

By addressing these concerns, India can create a research ecosystem where private funding drives innovation without compromising ethical standards or scientific freedom.

8.5 Government Policies to Regulate and Support Private Funding for Independent Research

As private-sector funding plays an increasingly critical role in research and innovation, governments must ensure that ethical, transparent, and unbiased funding mechanisms are in place. Without proper regulation, private funding can lead to corporate manipulation of research, bias in scientific inquiry, and neglect of fundamental research in favor of commercially viable projects.

To strike a balance between commercial interests and scientific progress, governments must implement strategic policies that encourage ethical private-sector funding, regulate corporate involvement, and promote public-private partnerships.

This section explores how tax incentives, regulatory frameworks, and public-private collaborations can support independent researchers while safeguarding scientific integrity.

8.5.1 Creating Tax Incentives for Ethical Private-Sector Funding of Research

One of the most effective ways to encourage private-sector investment in research is through tax incentives. Many countries, including the United States, Germany, and Japan, offer significant tax breaks to corporations, philanthropic organizations, and venture capitalists that fund scientific research. India can adopt a similar approach to ensure steady financial support for independent researchers.

Why Tax Incentives Are Necessary

- Encourages long-term private investment in R&D by reducing financial risks for corporations and philanthropists.
- Bridges the funding gap for independent researchers who struggle to access government grants.
- Promotes ethical funding practices by rewarding companies that invest in open-access, unbiased research.

Proposed Tax Incentive Models for India

1. Research and Development (R&D) Tax Credit for Private Funders

- Introduce a 150–200% tax deduction on investments made by corporations and philanthropic organizations in independent research.
- Provide additional tax credits for funding in critical areas such as renewable energy, healthcare, artificial intelligence, and fundamental science.

2. Patent-Linked Tax Benefits

- Companies funding independent research that leads to patents should receive reduced corporate tax rates on revenues generated from those patents.
- Similar to the UK Patent Box Model, where businesses pay lower tax rates on profits derived from patented innovations.

3. Exempting Research Grants from Income Tax

- Any research grants provided by private institutions, startups, or philanthropic organizations should be tax-exempt for independent researchers.

- This ensures that funding goes directly into research rather than being taxed as personal income.

4. Corporate Social Responsibility (CSR) Inclusion for Research Funding

- Modify India's CSR laws to include scientific research funding as an eligible activity.
- Require large corporations to allocate a percentage of their CSR funds to independent research projects.

By offering these tax incentives, the government can mobilize private capital for independent research, ensuring that non-affiliated scientists have access to consistent funding sources.

8.5.2 Regulating Corporate Involvement to Ensure Unbiased Scientific Inquiry

While corporate funding is essential for advancing research, it also carries significant risks of bias, suppression of unfavorable results, and commercial influence on scientific outcomes. Without proper regulations, companies might prioritize market-driven projects over fundamental research, leading to misleading findings and ethical violations.

Challenges in Corporate-Funded Research

- Selective Funding: Corporations may fund only projects that align with their business interests, neglecting independent or critical research.
- Manipulation of Results: Companies might pressure researchers to modify findings to favor their products or policies.
- Suppression of Negative Outcomes: If research exposes risks in corporate products, companies may block publication or

delay public access.

Proposed Government Regulations to Ensure Ethical Research
1. Mandatory Disclosure of Funding Sources

- Require all privately funded research projects to disclose their funding sources in publications and patents.
- Create a publicly accessible database listing corporate-backed research projects and their findings.

2. Establishing an Independent Ethics Review Board

Introduce a National Research Ethics Oversight Committee to evaluate corporate-funded research.

The committee should:

- Review corporate-funded research proposals before they receive funding.
- Ensure scientific integrity by preventing conflicts of interest.
- Audit research findings to detect and prevent manipulated or biased results.

3. Enforcing Open-Access Publication of Results

- Companies funding research should be legally required to publish all results, including negative findings.
- Prevents corporations from suppressing studies that could impact their commercial interests.

4. Whistleblower Protections for Researchers

- Strengthen legal protections for scientists exposing unethical corporate influence on research.
- Establish an anonymous reporting mechanism for researchers

facing pressure to modify findings.

By implementing these regulations, India can ensure that corporate-funded research remains independent, transparent, and free from commercial bias.

8.5.3 Introducing Public-Private Partnerships to Balance Commercial and Fundamental Research Goals

Public-Private Partnerships (PPPs) have successfully bridged funding gaps in scientific research in countries like Germany, the USA, and China. By creating structured collaborations between government agencies, private corporations, and independent researchers, India can ensure that both commercial and fundamental research receive adequate funding.

Benefits of Public-Private Research Collaborations

- Encourages private investment in long-term research projects that may not have immediate commercial applications.
- Allows independent researchers to access corporate R&D facilities without compromising scientific integrity.
- Balances innovation-driven market research with fundamental science to ensure sustainable scientific progress.

Proposed Public-Private Research Collaboration Models

1. Establishing a National Public-Private Research Fund

Create a government-backed research fund where:

- Corporations contribute a fixed percentage of their R&D budgets.
- The government matches private investments to encourage long-term research.
- Independent researchers can apply for funding without institutional affiliation.

2. Developing Industry-Academia-Independent Researcher Programs

Create structured research collaboration programs that allow:

- Independent researchers to work alongside corporate R&D teams.
- Government agencies to monitor and regulate industry-driven research projects.
- Tax incentives for companies funding both commercial and fundamental research projects.

3. Building Research Hubs with Shared Corporate and Government Infrastructure

Establish national research hubs where independent researchers can:

- Access state-of-the-art laboratories, supercomputing resources, and testing facilities.
- Collaborate with corporate-funded and government-backed projects.
- Ensure that research is conducted without institutional bias or corporate interference.

4. Expanding the 'Make in India' and 'Startup India' Initiatives to Support Research

Modify Make in India and Startup India policies to provide:

- Direct research funding for independent scientists working on deep-tech and emerging fields.
- Venture capital support for independent researchers seeking to commercialize scientific discoveries.
- Joint government-corporate sponsorships for long-term innovation projects.

By implementing structured public-private partnerships, India can create a balanced research ecosystem where both commercial innovation and fundamental scientific inquiry receive adequate support.

Conclusion

To ensure private-sector research funding benefits both scientific progress and ethical integrity, the Indian government must:

- Introduce tax incentives to encourage corporations and philanthropists to invest in independent research.
- Regulate corporate involvement to prevent research bias, manipulation, or suppression of negative findings.
- Establish structured public-private partnerships that promote both commercial innovation and fundamental science.

By adopting these policies, India can create a robust, independent research ecosystem that is transparent, ethically sound, and free from undue corporate influence, allowing independent researchers to thrive in a competitive, innovation-driven world.

9. Conclusion: The Future of Independent Research in India

India stands at a critical juncture in its scientific and technological evolution. While the country has made significant strides in industry, space exploration, pharmaceuticals, and digital innovation, its research ecosystem remains heavily dependent on institutional frameworks, bureaucratic funding mechanisms, and industry-driven priorities. In this environment, independent researchers—those who operate outside traditional institutions—face immense challenges in securing funding, accessing infrastructure, and gaining recognition for their work.

Throughout this book, we have examined the systemic barriers that hinder independent research and explored alternative funding mechanisms, policy recommendations, and successful global models that could help India build a more inclusive, dynamic, and innovation-driven research landscape.

In this final chapter, we summarize the dangers of ignoring independent innovation, present a call to action for policymakers, industries, and researchers, and outline key steps to create a sustainable research ecosystem in India.

9.1 The Danger of Ignoring Independent Innovation

If India continues to marginalize independent researchers and over-centralize scientific innovation within government institutions and elite universities, it risks:

1. Slowing Down Innovation and Global Competitiveness

- Many breakthrough discoveries in history have emerged from independent minds working outside traditional institutions (e.g., Einstein's theory of relativity, Tesla's electrical advancements, Ramanujan's mathematical insights).
- Ignoring independent researchers means losing out on unconventional ideas that could shape the next wave of scientific progress.

2. Increased Brain Drain

- Talented Indian researchers who fail to find funding, lab access, or recognition in India will continue to migrate to countries that support independent research, such as the US, Germany, China, and Japan.
- This results in India losing its best minds, forcing the country to import technologies instead of leading global innovation.

3. Over-Reliance on Institutional and Corporate Research Priorities

- A system that only funds institution-affiliated research limits innovation to bureaucratic agendas, short-term industry needs, and politically driven research projects.
- Without independent research, there is no intellectual diversity, no disruptive thinking, and no true progress.

4. Weak Global Research Presence

- Countries like the US and Germany consistently dominate global research rankings because they fund independent scholars, encourage open science, and promote entrepreneurial research ecosystems.
- Without structured support for independent research, India will struggle to compete with leading nations in AI, space technology, biotechnology, and advanced materials science.

Ignoring independent researchers means ignoring untapped potential—a mistake India cannot afford if it aims to become a global leader in science, technology, and innovation.

9.2 A Call to Action for Policymakers, Industries, and Researchers

To reverse these negative trends, stakeholders across government, industry, academia, and research communities must come together to support independent research as a national priority.

For Policymakers: Building a Supportive Ecosystem

- Expand research funding programs to include independent researchers without requiring institutional affiliation.
- Reduce bureaucratic barriers in research grant applications and streamline approval processes.
- Create a National Independent Research Council (NIRC) to oversee funding, ethics, and collaboration opportunities for non-affiliated scientists.
- Mandate open-access publishing for all publicly funded research, ensuring transparency and accessibility.
- Introduce tax incentives for corporations and philanthropists who fund independent researchers.

For Industry Leaders: Encouraging Private-Sector Support for Research

- Develop corporate-backed research grants that focus on both fundamental and applied sciences.
- Establish startup incubators that provide independent researchers with lab space, funding, and mentorship.
- Partner with government initiatives to co-fund high-impact research projects outside traditional institutions.
- Ensure ethical transparency in privately funded research to prevent corporate bias.

For Independent Researchers: Adapting to a Changing Research Landscape

- Leverage global funding opportunities, including crowdfunding, private grants, and international fellowships.
- Engage with industry collaborations and innovation hubs to access research facilities.
- Use digital platforms to publish and promote research, such as preprint servers (arXiv, bioRxiv) and open-access journals.
- Form independent research collectives to pool resources and enhance credibility.
- Advocate for policy changes by working with academic societies, think tanks, and policy forums.

Without proactive engagement from all stakeholders, India risks falling behind in the global race for scientific and technological leadership.

9.3 Steps to Create a Sustainable Research Ecosystem

A sustainable independent research ecosystem requires a long-term vision, systematic policy implementation, and collaborative engagement between the government, private sector, and research community.

Step 1: Institutionalizing Independent Research Funding

- Legislate a fixed percentage of national R&D funding to be allocated to independent researchers.
- Establish public-private research endowments that provide continuous financial support.

Step 2: Ensuring Open-Access Infrastructure for Independent Researchers

- Develop national research hubs with shared lab spaces and equipment for non-affiliated scientists.
- Expand university collaboration programs that allow independent researchers to access library resources, scientific journals, and technical databases.

Step 3: Strengthening Industry-Academia-Independent Researcher Partnerships

- Encourage industries to invest in open research projects that benefit both fundamental and applied science.
- Provide venture capital incentives for deep-tech startups led by independent scientists.

Step 4: Reforming Academic Publishing to Include Independent Researchers

- Abolish institutional affiliation requirements for publishing in high-impact journals.
- Reduce publishing costs for independent scholars through government subsidies and open-access initiatives.

Step 5: Internationalizing India's Research Ecosystem

- Develop global collaboration programs where independent Indian researchers can work with leading international institutions.
- Establish international research exchange grants specifically for non-affiliated scientists.

A thriving research ecosystem does not rely only on elite institutions—it empowers every capable mind to contribute to knowledge creation and innovation.

Final Thoughts: Embracing an Open and Inclusive Research Future

India's future as a knowledge-driven economy depends on its ability to foster a culture of independent inquiry, open science, and unrestricted research innovation.

By implementing policy reforms, industry collaborations, and inclusive funding models, India can:

- Unlock the potential of thousands of independent researchers.
- Encourage homegrown scientific innovation instead of relying on foreign technology.
- Attract global talent and position India as a world leader in scientific discovery.

The choice is clear—India can either continue with outdated, restrictive research models or embrace a bold, inclusive, and innovation-driven future.

The time to act is now.

If India truly aspires to be a scientific superpower, it must prioritize independent research as a national mission, ensuring that every researcher—regardless of institutional affiliation—has the opportunity to innovate, explore, and contribute to global progress.

Acknowledgments

Writing this book has been both a challenge and a necessity—a challenge because of the systemic barriers faced by independent researchers, and a necessity because their struggles deserve to be acknowledged and addressed. This book is not just a critique but also a call for change, and I could not have undertaken this journey alone.

Recognizing Mentors, Peers, and Contributors

I extend my deepest gratitude to mentors, educators, and thought leaders who have inspired my understanding of scientific inquiry, research ecosystems, and policy frameworks. Their insights have been instrumental in shaping the core arguments of this book.

I am also grateful to my peers and colleagues—both within academic circles and outside them—who provided critical feedback, shared personal experiences, and helped refine the ideas presented here. Their discussions and debates pushed me to examine the issue from multiple perspectives, ensuring that this book remains comprehensive and solution-driven.

To those who shared case studies, reports, and real-world examples, your contributions have strengthened the book's foundation, making the arguments more grounded in reality rather than theoretical speculation.

Acknowledging Independent Researchers Who Inspired This Book

This book is, above all, a tribute to independent researchers—those who pursue knowledge despite financial struggles, lack of institutional backing, and systemic neglect. Their persistence, resilience, and intellectual courage serve as a reminder that true scientific progress cannot and should not be confined to universities, corporations, or government agencies alone.

Many of the ideas presented in this book have been inspired by the voices of independent scholars who continue to innovate, experiment, and push the boundaries of knowledge despite having no structured support system. Their determination to challenge the status quo, seek truth beyond bureaucratic approvals, and contribute to global knowledge is what fuels the vision of this book.

Finally, to every researcher who has been denied funding, every scientist who has had to self-finance their discoveries, and every thinker whose ideas were rejected because they did not fit within institutional frameworks—this book is dedicated to you. May your contributions to science, technology, and humanity never go unnoticed.

Appendices

The following appendices provide practical resources and insights for independent researchers, policymakers, and industry stakeholders. These sections outline available funding opportunities, global best practices, and actionable steps for independent researchers in India.

Appendix A: List of Key Research Funding Agencies in India and Their Eligibility Criteria

1. Government-Funded Research Bodies

Agency	Focus Areas	Eligibility Criteria	Website
Science and Engineering Research Board (SERB)	Engineering, Physical Sciences, Life Sciences	Preference for university-affiliated researchers; some fellowships open to postdocs	www.serb.gov.in
Department of Science & Technology (DST)	Innovation, Technology, AI, Space, Environment	Primarily for institutional researchers, limited independent fellowships	www.dst.gov.in
Department of Biotechnology (DBT)	Biotechnology, Genetics, Medical Research	Preference for institutional partnerships; limited grants for independent researchers	www.dbtindia.gov.in
Indian Council of Medical Research (ICMR)	Public Health, Pharmaceuticals, Biomedical Research	Strong institutional affiliation required; fellowships for young scientists	www.icmr.nic.in
Council of Scientific and Industrial Research (CSIR)	Industrial R&D, Material Science, Pharmaceuticals	Institutional tie-up usually required; some projects allow individual applications	www.csir.res.in

2. Private and Philanthropic Research Grants in India

Organization	Focus Areas	Eligibility
Tata Trusts Research Grants	Social Sciences, Technology, Healthcare	Open to independent researchers, but institutional support is preferred
Infosys Science Foundation	Fundamental Sciences, Engineering, AI	Recognizes excellence but does not offer direct research grants
Wellcome Trust (India Alliance)	Biomedical Research, Public Health	Fellowships for early-career researchers; preference for institutional affiliation

Many of these funding bodies favor university-affiliated researchers, making it difficult for independent scholars to access direct funding.

However, some fellowships and international collaborations offer flexibility, which independent researchers in India should explore.

Appendix B: A Comparison of Global Independent Research Support Models

The following table compares how different countries support independent researchers, highlighting best practices that India could adopt.

Country	Key Research Funding Bodies	Support for Independent Researchers	Notable Policies
United States	National Science Foundation (NSF), Howard Hughes Medical Institute (HHMI)	Grants open to individual researchers, including postdocs and early-career scientists	Strong emphasis on independent fellowships
Germany	Max Planck Society, German Research Foundation (DFG)	Fully funded independent research fellowships	Open-access labs and collaboration networks
China	Thousand Talents Plan, National Natural Science Foundation of China (NSFC)	Direct state-funded grants for independent researchers	High state involvement in research funding
United Kingdom	Wellcome Trust, UK Research and Innovation (UKRI)	Fellowships and grants for independent researchers in science and humanities	Strong partnerships between universities and private funders
India	SERB, DST, CSIR	Limited opportunities for independent researchers	Research funding is largely tied to institutions

Key Takeaways for India

1. Increase direct-to-researcher funding opportunities similar to the US and Germany.
2. Create open-access lab networks so that independent researchers can use state-funded facilities.
3. Encourage corporate-backed research funding similar to

China's Thousand Talents Plan.

By implementing global best practices, India can expand research opportunities for non-affiliated scientists and prevent brain drain to other countries.

Appendix C: A Roadmap for Independent Researchers in India

For independent researchers struggling to navigate funding, resources, and publication, the following roadmap provides actionable steps to build a sustainable research career.

Step 1: Identifying Research Funding Sources

- Explore government grants that allow individual applications, such as SERB Early Career Research Awards.
- Apply for international fellowships, including Humboldt Research Fellowship (Germany) and Fulbright-Nehru Fellowship (USA).
- Consider private funding bodies, such as Tata Trusts and Wellcome Trust India Alliance.

Step 2: Accessing Research Facilities and Infrastructure

- Collaborate with open research labs, such as the National Supercomputing Mission (NSM) in India.
- Partner with innovation hubs and startup incubators to access state-of-the-art equipment.
- Apply for corporate research partnerships, which sometimes provide lab access for promising projects.

Step 3: Publishing and Gaining Recognition

- Use open-access preprint servers, such as arXiv, bioRxiv, or IndiaRxiv, to share research before formal publication.
- Submit papers to international open-access journals, reducing dependence on institutional backing.
- Participate in global research forums and conferences, even if presenting virtually.

Step 4: Networking and Collaboration

- Join online research communities like ResearchGate and Academia.edu to connect with international scholars.
- Seek mentorship through professional associations, such as IEEE, ACM, or the Indian Science Congress.
- Collaborate with corporate R&D teams for applied research projects.

Step 5: Exploring Alternative Funding Models

- Use crowdfunding platforms like Experiment.com for niche projects.
- Apply for venture capital or angel funding if research has commercial potential.
- Leverage incubator and accelerator programs that support research-driven startups.

Step 6: Advocacy and Policy Engagement

- Engage with science policy forums to push for research-friendly policies.
- Collaborate with NGOs and think tanks advocating for better independent research funding.

- Raise awareness through media, podcasts, and blogs about the challenges of independent research.

By following this roadmap, independent researchers in India can navigate systemic barriers, build credibility, and secure funding opportunities despite existing challenges.

Final Thoughts on the Appendices

These appendices aim to provide practical guidance for independent researchers and policymakers. India has the talent and intellectual capacity to lead in global research, but this will only be possible if barriers to independent research are dismantled and funding mechanisms become more inclusive.

Independent researchers should not be forced to leave India in search of better opportunities—instead, India should create an ecosystem where innovation thrives beyond institutional boundaries.

Glossary

This glossary provides concise definitions of key terms used throughout the book, helping readers better understand research funding mechanisms, institutional structures, and policy-related terminology.

A

Academic Publishing Bias

The tendency of journals and conferences to prioritize research from prestigious institutions, making it harder for independent researchers to get published.

Angel Investor

A private individual who provides financial backing to early-stage startups, innovation projects, or research-driven ventures in exchange for equity or future returns.

B

Brain Drain

The migration of highly skilled researchers and scientists from one country to another due to better funding, research infrastructure, and career opportunities abroad.

Bureaucratic Hurdles

Delays and inefficiencies in research funding approvals due to excessive paperwork, slow decision-making, and rigid institutional processes.

C

Corporate-Backed Research

Scientific studies and innovation projects funded by private corporations, often with an emphasis on market-driven outcomes rather than fundamental scientific inquiry.

Crowdfunding for Research

A method where researchers raise funds from the public via online platforms like Experiment.com, Kickstarter, and GoFundMe to support independent scientific studies.

D

Disruptive Innovation

Scientific breakthroughs or technological advancements that fundamentally change existing industries, markets, or scientific paradigms.

Direct Research Grants

Funding provided directly to researchers without requiring them to be affiliated with an institution. Such grants are rare in India but common in countries like the US and Germany.

F

Fundamental Research

Basic scientific investigation aimed at expanding knowledge without immediate commercial applications. Often underfunded compared to applied research.

Fellowship for Independent Researchers

A financial award provided to individual researchers for conducting their studies without needing institutional affiliation. Examples include the Humboldt Fellowship (Germany) and Wellcome Trust Fellowships (UK/India).

G

Grant Disbursement Process

The steps involved in applying for, reviewing, approving, and receiving research funding. Many grant systems favor institution-affiliated researchers over independent scholars.

Global Research Collaboration

Partnerships between researchers from different countries that promote knowledge exchange, funding opportunities, and access to international labs.

I

Incubators and Accelerators

Organizations that support research-driven startups by providing funding, lab space, mentorship, and commercialization opportunities. Examples include IIT Bombay's SINE and Y Combinator.

Independent Researcher

A scientist or academic who conducts research without being formally affiliated with a university, government lab, or corporate R&D unit.

Institutional Monopoly on Research Funding

The dominance of universities, IITs, and government-backed labs in receiving the majority of public research funding, making it difficult for independent researchers to secure grants.

O

Open-Access Publishing

A publishing model where scientific papers are freely available to the public without paywalls, promoting greater accessibility and knowledge sharing.

Open Science Movement

A global initiative that encourages collaborative, transparent, and publicly accessible research, reducing barriers for independent scholars.

P

Peer Review System

The process by which research papers are evaluated by other experts in the field before publication. Often favors institutional researchers due to existing academic networks.

Predatory Journals

Low-quality, pay-to-publish journals that exploit researchers by charging high fees without conducting rigorous peer review.

Public-Private Partnership (PPP) in Research

A collaborative funding model where government agencies and private corporations co-fund research projects, ensuring a balance between commercial and fundamental research.

R

Research Commercialization

The process of turning scientific discoveries into marketable products or technologies, often prioritized in corporate-funded research.

Research Ethics and Integrity

Standards ensuring that scientific studies are conducted transparently, without bias, conflicts of interest, or corporate manipulation.

S

Startup-Led Research

Scientific studies and technological advancements funded by startups, particularly in AI, biotech, and space exploration.

Scientific Gatekeeping

The practice where elite institutions and academic journals control access to research funding, publishing opportunities, and professional networks, making it harder for independent researchers to gain recognition.

T

Tax Incentives for Research

Financial benefits given by governments to corporations, philanthropists, and private entities that invest in research and development. Countries like Germany, the US, and China use tax breaks to encourage private-sector research funding.

V

Venture Capital (VC) for Research Startups

Investment funding from VC firms or angel investors that helps independent researchers commercialize their discoveries by launching startups.

Final Thoughts on the Glossary

This glossary provides a foundational understanding of the key issues surrounding independent research, funding barriers, and policy-driven solutions. As India works toward building a more inclusive research ecosystem, these terms will remain critical to shaping future discussions on innovation, funding equity, and scientific progress.

References & Bibliography

This section includes reports, policy documents, academic studies, and news articles that have informed the analysis in this book. Reliable data, historical case studies, and global research trends have been used to highlight the challenges faced by independent researchers in India and to propose practical solutions based on global best practices.

1. Government Reports and Policy Documents

- Science, Technology and Innovation Policy (STIP 2020) – Government of India, Department of Science & Technology.
- Research and Development Statistics (2022) – Ministry of Science and Technology, Government of India.
- Higher Education in India: Vision 2030 – FICCI and Ernst & Young Report, 2013.
- Annual Report of the Department of Science & Technology (DST), India – Government of India.
- Global Innovation Index Report (2023) – World Intellectual Property Organization (WIPO).
- UNESCO Science Report (2021): The Race Against Time for Smarter Development – UNESCO.
- India's R&D Expenditure Trends (NITI Aayog, 2022) – Government of India.
- CSIR Annual Research Report (2023) – Council of Scientific

and Industrial Research (CSIR), India.

- National Education Policy (NEP 2020) – Ministry of Education, Government of India.
- Startup India: The Innovation Policy Initiative – Government of India, 2016.

2. Academic Studies and Research Papers

- Abrol, D. (2014). *Research and Development in India: The Institutional Framework and Policy Dynamics.* Indian Journal of Science and Technology, 7(10), 1602–1613.
- Gupta, R. & Sharma, M. (2020). *Challenges and Opportunities for Independent Researchers in India.* Journal of Indian Science Policy, 15(2), 87–102.
- Kumar, A. (2021). *The Impact of Bureaucratic Hurdles on Scientific Innovation in India.* International Journal of Public Policy Research, 10(3), 211–230.
- Das, P. & Bhattacharya, S. (2018). *The Role of Private Funding in Research and Development: Lessons from Global Models.* Indian Journal of Policy Studies, 12(1), 45–63.
- Banerjee, S. (2022). *The Brain Drain Effect: How Indian Researchers Seek Better Opportunities Abroad.* Journal of Global Mobility Studies, 9(2), 133–150.
- Choudhury, V. (2019). *Public vs. Private Research Funding: A Comparative Study of India, China, and the US.* Global Science Review, 14(4), 98–117.

3. News Articles and Reports on Research Trends

- "Why India's Research Investment Is Below Global Standards" – The Hindu, 2023.
- "How Indian Scientists Are Turning to Crowdfunding for

Research" – The Indian Express, 2022.

- "India's Research Output Lags Behind China and the US" – Times of India, 2021.
- "Private Sector Must Step Up R&D Funding: NITI Aayog Report" – Economic Times, 2022.
- "The Struggles of Independent Researchers in India" – Hindustan Times, 2023.
- "Corporate Influence on Scientific Research: A Growing Concern?" – Forbes India, 2022.
- "India's Science Budget: Why It's Not Enough for a Knowledge Economy" – Financial Express, 2023.
- "Independent Researchers Face an Uphill Battle: No Labs, No Funding" – BBC India, 2021.

4. Case Studies Referenced in the Book

A. Global Research Funding Models

- Howard Hughes Medical Institute (HHMI), USA – Funding independent biomedical researchers.
- European Research Council (ERC) Grants – Supporting individual excellence in frontier research.
- Max Planck Society, Germany – Open-access research and independent fellowships.
- Thousand Talents Plan, China – Attracting and funding top independent researchers globally.
- NASA's Public-Private Research Collaborations – How private funding and government cooperation foster innovation.

B. Independent Research Success Stories

- Srinivasa Ramanujan – A self-taught mathematician who

contributed to number theory despite lacking institutional support.

- Jagadish Chandra Bose – Conducted pioneering research in wireless communication with minimal funding.
- Elon Musk's SpaceX Model – How private funding reshaped aerospace innovation.
- DeepMind (UK) – AI research startup acquired by Google, demonstrating the power of independent AI innovation.

C. Indian Research Challenges and Funding Trends

- The ISRO Model – How India's space agency succeeded despite limited initial funding.
- IITs and Exclusive Research Monopolies – How funding is concentrated in elite institutions.
- Pharmaceutical Industry's Weak R&D Investment – Why India excels in generic drugs but lags in innovation.
- The Semiconductor Industry Missed Opportunity – Why India failed to develop a domestic chip manufacturing ecosystem.

Final Thoughts on the Bibliography

The references listed here serve as a foundation for understanding the research funding landscape in India. They highlight the structural barriers, global best practices, and case studies that illustrate both challenges and potential solutions.

For readers who want to explore these topics further, these resources provide deeper insights into how independent research can be supported and integrated into India's broader innovation strategy.

Call to Action: What Readers Can Do

Scientific progress thrives when ideas are nurtured, independent thinkers are supported, and innovation is not constrained by bureaucratic red tape. Throughout this book, we have explored the systemic barriers facing independent researchers in India, from institutional monopolies and funding restrictions to industry neglect and publication biases.

However, change is possible—but it requires collective action from policymakers, industry leaders, and individual researchers. This section outlines practical steps that readers can take to help create a more inclusive and dynamic research ecosystem in India.

5.1 For Policymakers: Reforming Research Policies for Inclusivity

1. Create Direct Funding Mechanisms for Independent Researchers

- Introduce grants that do not require institutional affiliation, similar to the European Research Council (ERC) model.
- Simplify the application process for research grants, removing excessive bureaucratic hurdles.
- Establish a National Independent Research Council (NIRC) to oversee funding, mentorship, and open-access research initiatives.

2. Increase Government Investment in R&D

- Raise India's R&D spending from the current ~0.7% of GDP to at least 2%, aligning with global innovation leaders.
- Allocate a percentage of science and technology funding specifically for non-institutional researchers.

3. Open Labs and Resources to Independent Researchers

- Make state-funded research labs, libraries, and databases accessible to non-affiliated scientists.
- Implement an open-access research repository, where independent scholars can publish and access research without paywalls.

4. Strengthen Public-Private Partnerships in Research

- Introduce tax incentives for corporations investing in independent research.
- Encourage startup incubators to fund and mentor research-driven projects.
- Develop collaborative research hubs, where independent researchers, universities, and industries can work together.

Policymakers can initiate these changes by advocating within government bodies, submitting policy recommendations, and engaging with researchers to understand their needs.

5.2 For Industry Leaders: Supporting Research Beyond Universities

1. Fund Independent Research Through Private Grants

- Establish corporate-backed research fellowships for non-affiliated scientists.

- Invest in open innovation projects, where research findings are shared publicly instead of being patented for exclusive corporate use.

2. Build Research Incubators and Innovation Hubs

- Develop corporate-funded research hubs, where independent researchers can access lab equipment, mentorship, and industry collaborations.
- Support deep-tech startups led by independent scientists, similar to the Y Combinator model.

3. Promote Open Science and Research Ethics

- Adopt transparent research funding policies, ensuring that privately funded studies maintain scientific integrity.
- Ensure that corporate-funded research does not manipulate outcomes for commercial interests.

Industry leaders can support these initiatives by partnering with independent researchers, funding high-impact projects, and fostering a culture of innovation within their organizations.

5.3 For Independent Researchers: Overcoming Barriers and Advocating for Change

1. Explore Alternative Funding Sources

- Apply for international research fellowships, such as the Humboldt Fellowship (Germany) and Wellcome Trust Grants (UK/India).
- Consider crowdfunding platforms like Experiment.com for niche projects.
- Seek venture capital or angel investment if research has commercial potential.

2. Leverage Open-Access Platforms and Global Networks

- Publish research on preprint servers (arXiv, bioRxiv, IndiaRxiv) to gain visibility.
- Collaborate with global researchers through platforms like ResearchGate and Academia.edu.
- Contribute to open-access journals instead of paywalled publications.

3. Join Advocacy Groups and Push for Policy Reform

- Engage with science policy forums and professional associations to demand research-friendly policies.
- Support petitions and public campaigns for increased research funding.
- Use social media and blogs to raise awareness about the challenges faced by independent researchers.

Independent researchers can take charge of their careers by networking strategically, utilizing global resources, and actively advocating for fair research policies.

5.4 How Every Reader Can Support Independent Research

Even if you are not a policymaker, industry leader, or researcher, you can still play a vital role in supporting independent science. Here's how:

1. Support Open Science Initiatives

- Sign petitions advocating for government-backed open-access journals.
- Donate to research crowdfunding campaigns that support independent scientists.
- Follow and amplify the work of independent researchers

through social media, blogs, and academic platforms.

2. Encourage Public Engagement with Science

- Support science communication initiatives that make research accessible to the general public.
- Attend public science lectures, TED Talks, and knowledge-sharing events.
- Encourage educational institutions to introduce programs that fund non-traditional researchers.

3. Demand Greater Accountability in Research Funding

- Write to local policymakers and science ministries, urging them to prioritize independent research funding.
- Hold universities and industries accountable for excluding independent scholars from funding opportunities.

Every reader can be an advocate for change by raising awareness, supporting open-access research, and pressuring decision-makers to act.

Final Thoughts: The Power of Collective Action

Change does not happen in isolation. If policymakers, industry leaders, and researchers work together, India can transform its research ecosystem into one that:

- Funds innovation without institutional bias
- Encourages open collaboration between academia, industry, and independent researchers
- Eliminates bureaucratic obstacles to scientific discovery
- Creates world-class research opportunities at home, reducing brain drain

This book is a starting point, but real change requires action. The future of independent research in India depends on collective advocacy, policy reforms, and industry support.

The time to act is now. Join the movement, demand change, and be a part of shaping India's next scientific revolution.

Author's Note

The journey of writing this book has been deeply personal and intellectually challenging. As an independent researcher in Physics and Universal Theories, I have firsthand experience of the barriers that non-affiliated scientists face—the lack of funding, restricted access to research infrastructure, and the systemic bias that favors institutional researchers over independent thinkers.

This book was not written just as a critique but as a call for reform—a way to highlight the urgent need for inclusivity in research funding, open-access scientific collaboration, and policy changes that recognize the value of independent innovation. India has no shortage of talent, yet the current system marginalizes researchers who do not conform to traditional academic structures, leading to a wasted pool of intellectual potential and brain drain.

Why This Book Was Written

Scientific progress has never been confined to institutions alone. Some of history's greatest thinkers—Einstein, Tesla, Ramanujan, and Bose—pursued their work independently, often against great odds. Today, however, the research ecosystem is designed to exclude those who do not fit within the rigid frameworks of universities, corporations, or government-backed labs.

I wrote this book to expose these barriers, propose solutions, and initiate a conversation about the future of research in India. Independent researchers, whether in physics, engineering, artificial intelligence, or medical sciences, should not have to struggle for basic resources. Knowledge should not be monopolized by a handful of institutions; it should be open to all those with the passion and capability to contribute.

Through this book, I hope to:

- Raise awareness about the systemic exclusion of independent researchers.
- Encourage policy discussions that promote open-access research and direct funding for non-affiliated scientists.
- Provide alternative pathways for independent researchers to secure funding, publish their work, and gain recognition.
- Highlight global models that India can learn from to create a more inclusive research environment.

What Needs to Change in the Next Decade

If India aspires to be a global leader in science and technology, the research ecosystem must undergo fundamental reforms. The following key areas require immediate attention over the next decade:

1. Creating Direct Research Fellowships for Independent Scientists

- The government must introduce grants that do not require university affiliation, allowing researchers to work independently.
- Fellowships for non-institutional scientists should be established, similar to the European Research Council (ERC) model.

2. Open-Access Research and Laboratory Facilities

- National research infrastructure should be accessible to all—not just university-affiliated researchers.
- High-quality scientific journals should remove paywalls,

enabling independent researchers to publish and access knowledge freely.

3. Encouraging Private and Crowdsourced Research Funding

- Philanthropic organizations and private foundations must step up to support independent scientists.
- Crowdfunding models, similar to Experiment.com, should be promoted to fund cutting-edge projects.

4. Strengthening Industry Collaboration with Independent Researchers

- Corporations should fund research that extends beyond immediate business goals, supporting fundamental and theoretical science.
- Incubators and research hubs should integrate non-affiliated researchers, providing them with workspace, mentorship, and technical support.

5. Challenging the Institutional Monopoly on Research and Publishing

- High-impact journals should abolish institutional affiliation requirements, ensuring independent researchers have equal opportunities to publish.
- Peer review processes must be reformed to eliminate bias against non-affiliated scientists.

Final Thoughts: A Call for Scientific Freedom

The future of research should not be dictated by bureaucracy, institutional favoritism, or corporate agendas. It should be driven by curiosity, innovation, and the pursuit of knowledge for the betterment of society.

As an independent researcher, I know how frustrating and isolating this journey can be. But I also believe that change is possible—if we collectively demand it. This book is only the beginning of that conversation. The next step is to transform ideas into action, push for policy changes, and create an environment where knowledge is free, accessible, and inclusive.

To all independent researchers, thinkers, and innovators—I see you, I stand with you, and I hope this book contributes to making your journey a little easier.

Let's work together to redefine the future of scientific inquiry—one that belongs to everyone, not just the privileged few.

Sandeep Chavan

About the Author

Sandeep Chavan-Author | Researcher | Educator

Sandeep Chavan is an independent researcher, author, and educator with a deep interest in Physics, Universal Theories, Scientific Inquiry, and Policy Reforms. With an extensive background in engineering, education, and industrial research, he has dedicated his career to exploring complex scientific phenomena, analyzing systemic challenges in research ecosystems, and advocating for policy changes that support independent researchers.

Academic and Research Contributions

As an independent researcher, Sandeep has conducted in-depth studies on scientific methodologies, technological advancements, and the challenges faced by non-affiliated scientists. His research papers and articles have been published in open-access platforms, contributing to global discussions on knowledge accessibility, innovation, and research ethics.

Some of his published research can be accessed through:

Zenodo Research Papers[1]

Additional Publications on Zenodo[2]

Physics and Universal Theories Research[3]

Scientific Inquiry and Knowledge Expansion[4]

Books by Sandeep Chavan

A prolific writer, Sandeep has authored multiple books that explore scientific principles, human behavior, policy reforms, and India's socio-economic landscape. His works critically examine the intersection of technology, industry, and education, providing readers with well-researched insights and practical solutions.

His books are available on major platforms worldwide, including: Amazon, Barnes & Noble, Apple Books, Google Books, Hoopla, etc

Advocacy for Independent Research and Open Science

Throughout his career, Sandeep has been a vocal advocate for independent researchers, highlighting the challenges they face due to institutional barriers, funding monopolies, and limited access to research infrastructure. His work emphasizes the need for policy reforms, open-access publishing, and decentralized research ecosystems that empower scientists outside traditional academic frameworks.

Vision and Future Endeavors

As a researcher and author, Sandeep Chavan continues to challenge conventional knowledge structures and advocate for scientific inclusivity and equitable research opportunities. He remains committed to:

---

1. https://doi.org/10.5281/zenodo.14791053
2. https://doi.org/10.5281/zenodo.14788539
3. https://doi.org/10.5281/zenodo.14782900
4. https://doi.org/10.5281/zenodo.14793162

- Breaking barriers in independent research funding and accessibility.
- Promoting open science and ethical research frameworks.
- Exploring deeper connections between physics, philosophy, and universal theories.
- Bridging the gap between research policy and scientific progress.

Through his books, research, and advocacy, Sandeep Chavan strives to reshape how knowledge is created, shared, and applied for the betterment of society.

Further Reading

For readers who wish to explore deeper insights into science policy, research funding, innovation ecosystems, and the role of independent researchers, the following books, research papers, and reports provide valuable perspectives on global best practices and systemic challenges in research.

1. Books on Science Policy and Research Funding

The Science of Science Policy: A Handbook – Kaye Husbands Fealing, Julia Lane, John H. Marburger III

A comprehensive guide on how governments and institutions fund and evaluate research, and the impact of policy decisions on scientific progress.

The New Science of Strong Materials – J.E. Gordon

Explores how scientific discovery and innovation have shaped material sciences, with insights into independent research contributions.

The Structure of Scientific Revolutions – Thomas S. Kuhn

A must-read for researchers, explaining how scientific paradigms shift over time, often led by independent thinkers challenging mainstream theories.

Science Fictions: Exposing Fraud, Bias, Negligence and Hype in Science – Stuart Ritchie

Investigates the flaws in academic research, including peer review bias, grant funding manipulation, and institutional politics.

How Economics Shapes Science – Paula Stephan

A critical analysis of how financial incentives, institutional funding, and industry collaborations influence scientific research worldwide.

Open Science: Sharing Knowledge in the Global Century – Sönke Bartling, Sascha Friesike

Examines the future of open-access research and why traditional academic publishing models restrict knowledge dissemination.

2. Research Papers and Reports on Innovation and Independent Research

The Global Innovation Index Report (2023) – World Intellectual Property Organization (WIPO)

A detailed ranking of countries based on R&D investment, patent output, and scientific collaboration.

UNESCO Science Report (2021): The Race Against Time for Smarter Development – UNESCO

Covers global challenges in research funding and the rise of private sector involvement.

Scientific Research in India: The Institutional vs. Independent Debate – Journal of Indian Science Policy

Investigates why India's research funding is concentrated in elite institutions and how this affects scientific innovation.

Research and Development Expenditure Trends (2022) – NITI Aayog, Government of India

Highlights the low R&D spending in India compared to China, the US, and Germany, and suggests funding reforms.

Open Science and the Future of Knowledge – European Commission

Discusses how global research communities are moving toward open-access knowledge-sharing models.

Public vs. Private Research Funding: A Comparative Study of India, China, and the US – Global Science Review

Explores how private research grants and public-private partnerships are shaping the research ecosystem in leading economies.

Crowdfunding for Science: A Disruptive Alternative to Traditional Research Funding? – Science Advances

Analyzes the growing role of crowdfunding platforms in funding independent research.

The Role of Industry-Academia Collaboration in Scientific Innovation – Harvard Business Review

Examines how corporate partnerships influence the direction of scientific research and why independent research often gets overlooked.

3. Reports on Challenges in Publishing and Open-Access Movements

The Future of Academic Publishing: Rethinking Peer Review and Institutional Bias – MIT Press

Discusses how peer review systems are flawed and why independent researchers struggle to publish in high-impact journals.

The State of Open Data (2022) – Digital Science & Springer Nature

Examines how open-access publishing and open data initiatives are transforming global research.

Predatory Journals: A Threat to Scientific Integrity – Nature

Investigates the rise of pay-to-publish journals and how they exploit independent researchers.

Final Thoughts on Further Reading

The resources listed above provide invaluable insights into the global research landscape, funding challenges, and innovation policies. By reading and engaging with these materials, researchers, policymakers, and industry leaders can:

- Understand how research funding models impact scientific progress.
- Recognize the systemic barriers independent researchers face.
- Explore solutions for open-access science and funding equity.

These books, papers, and reports serve as a foundation for future discussions, helping to reshape research policies and create a more inclusive scientific environment for all.

Disclaimer

T***he Forgotten Innovators: Why India Fails Its Independent Researchers*** is a critical analysis of the challenges faced by non-affiliated researchers in India and globally. The views, opinions, and recommendations presented in this book are based on research, case studies, policy analysis, and personal experiences as an independent researcher.

This book is not intended to discredit any specific institution, organization, or government body, but rather to highlight systemic barriers in the research ecosystem that hinder scientific progress outside institutional frameworks. While every effort has been made to ensure accuracy and objectivity, the content should be interpreted as an informed critique rather than an absolute authority on research policy.

The book does not provide legal, financial, or professional advice regarding funding applications, institutional policies, or research grants. Readers are encouraged to consult official sources and experts when making decisions related to research funding, academic publishing, or policy advocacy.

The author and publisher do not assume any responsibility for actions taken based on the information provided in this book. The content is meant to inform, provoke discussion, and advocate for necessary reforms in the research landscape.

All references to government policies, industry practices, and funding mechanisms are accurate to the best of the author's knowledge at the time of writing, but research and funding landscapes continuously evolve. Readers are advised to verify data and consult official research bodies and funding agencies for up-to-date information.

Any resemblance to real persons, research projects, or organizations is purely coincidental unless explicitly stated. The use of case studies and historical examples is for illustrative purposes only and does not imply endorsement or criticism of any individual or entity.

By reading this book, the reader acknowledges and agrees that the author and publisher are not liable for any interpretations, actions, or consequences resulting from the material presented.

Don't miss out!

Visit the website below and you can sign up to receive emails whenever SANDEEP CHAVAN publishes a new book. There's no charge and no obligation.

https://books2read.com/r/B-A-EPGPC-JLEYF

BOOKS 2 READ

Connecting independent readers to independent writers.

Did you love *The Forgotten Innovators*? Then you should read *Human Side of Einstein*[1] by SANDEEP CHAVAN!

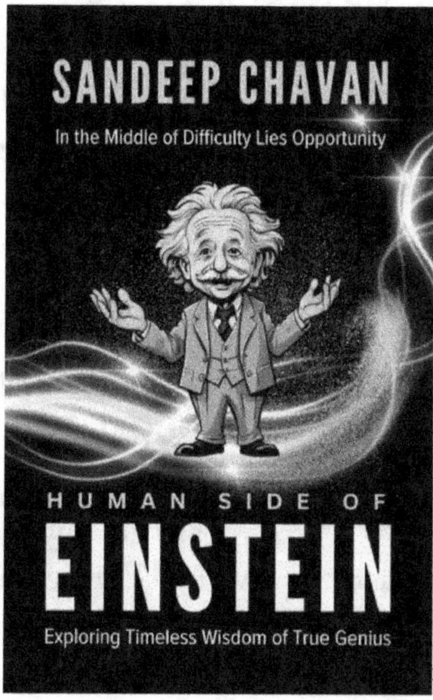

[2]

Human Side of Einstein: Exploring Timeless Wisdom of True Genius

Albert Einstein is celebrated as one of the greatest scientific minds in history, but his profound reflections on life, humanity, and purpose extend far beyond the confines of physics. In *Human Side of Einstein: Exploring Timeless Wisdom of True Genius*, Sandeep Chavan delves into Einstein's timeless words, offering readers a deeper understanding of his philosophies and their relevance to modern life.

This book is not a scientific treatise but an exploration of Einstein's humanistic insights. Each chapter unpacks one of his famous quotes, presenting it through the lenses of philosophy, psychology, and

1. https://books2read.com/u/4NNGyo

2. https://books2read.com/u/4NNGyo

practical living. From the transformative power of imagination to the enduring value of simplicity, the chapters examine universal themes that resonate with readers from all walks of life.

Chavan highlights Einstein's belief in selflessness, lifelong learning, and the responsibility of fostering peace and understanding in a divided world. The book addresses critical topics like the ethical use of technology, the importance of global collaboration, and the need for empathy in solving humanity's challenges. Through thoughtful reflections, real-world examples, and engaging exercises, readers are encouraged to apply Einstein's wisdom to their own lives, embracing a mindset of curiosity, resilience, and purpose.

The title, *Human Side of Einstein*, reflects the book's intent to go beyond the equations and theories to reveal a compassionate, curious, and profoundly human thinker. Einstein's assertion that "In the middle of difficulty lies opportunity" serves as a guiding principle throughout the book, inspiring readers to view life's challenges as opportunities for growth and transformation.

With a blend of philosophical depth and actionable insights, *Human Side of Einstein* invites readers to explore the essence of genius not as an extraordinary gift but as the capacity to think deeply, act meaningfully, and contribute to the betterment of the world. Whether you are a student, professional, or lifelong learner, this book offers timeless lessons to navigate an ever-changing world with clarity, integrity, and imagination.

Through Einstein's lens, Chavan presents a roadmap for living thoughtfully and purposefully, ensuring that the genius of one man becomes a guiding light for humanity's collective journey.

Read more at https://www.sandeepjchavan.com/.

Also by SANDEEP CHAVAN

Echoes of War: The Russia-Ukraine Conflict and Its Global Ripples
Shattered Lives
The War Economy
Borders Redrawn
Arms and Allies
Collateral Damage
Justice in the Ashes

Somewhere in the Darkness of the Cosmos
The Blueprint of the Universe

The Eternal Sage Chronicles
The Sage and the Sovereign
The Family Pillar
The People's Flame
The Warrior's Soul

The Fundamental Theory of Life & Existence
Energy in Action: The Soul Demystified
The Active Mind
Energy in Connection
Death: The Disconnection

Ultra Modern Life Philosophy
The Duality of Laughter and Tears
Compare to Accept

Standalone
It's Not AI, It's AHI - Amplified Human Intelligence
The Decision Paradox
The Four Sapiens
Malicious Script of Indian Polity
The IIT Legacy & Global Impact
The IIT Legacy & Global Impact
You are just a Version of Your Original
Win the Game You Didn't Choose
Always Keep Your Bags Packed
Reveal to Shield - Learn the Game of Facades
Engage Beyond Elections
Culture, Identity & Change: The Evolution of Indian Society
The Triangular Dynamics
Never Trust Without Doubt
If You Can't Change the Country, Leave the Country
Live Fearless with Fear
India: The Investment Magnet

Realignment
Powering a Billion Dreams
Universal Skill
Extension: From Brush to Algorithm
Human Side of Einstein
The Future Knows No Lines
Let's Bring the Generations to the Table
Unified Energy Dynamics
A Beat That Never Dies
Strategic Horizons
Right From Scratch
The Forgotten Innovators

Watch for more at https://www.sandeepjchavan.com/.